Muse of Fire

Storytelling & The Art of Science Communication

Muse of Fire

Storytelling & The Art of Science Communication

Tim Miller

SPOKEN SCIENCE PRESS

Spoken Science edition, 2015

www.spokenscience.com

ISBN 978-0-996-94920-0 (pbk.)

The first principle is that you must not fool yourself and you are the easiest person to fool.

-Richard Feynman

Contents

Foreword

As a young boy I dreamed of travelling to the stars, so when I got to university I decided to study them. Such dreams were not uncommon among American boys and girls of my generation, for we came of age in an era when space travel seemed not only possible, but inevitable. We had been fed a steady diet of the science fiction franchises that ruled popular culture, and the Apollo program seemed so present in the country's consciousness that we were scarcely aware it had ended before we were born. The highest praise a parent or teacher could give a child in those days was: "Someday he'll be an astronaut."

And so I decided to study the stars, not because I so desperately wanted to know more about how they worked, but because knowing more about them fed my ability to imagine being there.

After several years of diligent pursuit of my physics degree, because (I was surprised to learn) astronomy and cosmology are really branches of physics, I travelled in my senior year to a mountaintop in Arizona, where I and my advisor spent ten nights taking measurements of the cataclysmic binary stars that were his specialty. I was to operate a 1.3 meter telescope, tracking a handful of these stars, taking data.

I found this exercise surprisingly dull, for after opening the dome and filling the dewars with liquid nitrogen and typing your targets into the tracking software, there wasn't really a whole lot to do. Science, it seemed, involved lots of waiting, and I wasn't good at waiting.

There was another researcher who was with me in the control room most of those nights, and he seemed to love waiting. He would sit at the console, watching a stream of unprocessed data as it cascaded down the screen, utterly rapt, apparently seeing something I could not.

Cataclysmic binary stars are fascinating systems: they consist of a small but incredibly dense white dwarf star, closely orbited by another star quite like our sun. The stars are so close that the dwarf pulls material off of the second star, which accumulates in a disk, growing thicker and heavier, until periodically it ignites in runaway fusion, massively increasing the brightness of the system. The transfer of that mass from one star to the other was our primary area of interest.

During the wee hours of one of these nights, as I filled a page with doodles that were half equations and half cartoons, and my companion sat engrossed in his screen, I turned to him.

"Do you ever wonder if any of these stars have planets, and if there are people living on them? What would it be like to live on a world where the brightness of the sun varied from day to day, a world a where without warning the sky itself could ignite and char the planet to ash? Or a cold, distant world, a world of near-constant winter, except for every few years, when the sun would erupt, and a glorious summer would blossom? What would those people make of it all? Who would be their gods?"

Without looking away from his screen, my companion answered: "They'd sure have much better data on the mass exchange dynamics."

I think that was the moment I decided I wasn't going to be a scientist.

Though I went on to complete a graduate degree in engineering, there too I found that I lacked the passion for diligent, dogged pursuit of data that scientific research requires. My graduate advisors and I agreed: I was better at talking about science than I was at doing it.

Instead, I have managed to build a career around science communication. I've been a presenter of live shows at the Museum of Science in Boston, a writer for a think tank at the University of Southern California, a consultant to the Exploratorium in San Francisco, a lecturer at the Engineering School at Harvard University, and I'm pretty sure I have the peculiar distinction of being the only person in America ever paid by the

National Science Foundation to attend an art school. (It's a long story). I've written screenplays in Hollywood, appeared on television in New England, been a press officer at a major research university, and done many, many other things.

For much of the last decade I've travelled and lectured at universities and professional societies, both alone and in collaboration with other organizations, teaching working scientists how to communicate. This work has been tremendously rewarding, and has given me the unique opportunity to speak and work with scientists from a wide range of disciplines. What I have found is that the problems scientists face are the same regardless of their field. We all face similar challenges in communicating about our work, and there are (as you shall see) some simple principles I think we can use to overcome them.

There was a period of time in my life, before physics, before I went to university, long before that night at the telescope, where my great love was not science, but theater. I fancied myself an actor and budding playwright. My favorite part of working in the theater was the day after the run of a show, when the sets would be dismantled, the costumes and props packed away. I loved the quiet moments when the cast had gone home, and the stage was bare and black. There is something sweet and sad about an empty stage, but it always filled me with a sense of possibility. An empty stage could become anything, with a little paint and a dose of imagination, it could be the Amazon rainforest, the Sistine Chapel, or the surface of Mars.

The title of this book comes from the opening lines of Shakespeare's play Henry V, in which the chorus (in this case basically a narrator) asks the audience to suspend their disbelief. We cannot fill the air with arrows, or gallop the horses of an army onstage, he says, so let your imaginations bring them instead. (There's a great version of this introduction in Kenneth Branagh's 1989 film version of the play).

I chose this as the title for this book because over the years I've come to believe that what the narrator is asking for is the biggest thing missing from science communication. Scientists generally don't want to suspend their disbelief. They don't want to imagine things. That's obviously because science is based on disbelief, science forbids imagination. But science isn't human. Scientists are. And humans are storytellers.

I never really walked away from my first love when I turned to science, and over the years I've continued to pursue theater, film, and fiction in general alongside my career in science communication. I've had the chance to study playwriting, and screenwriting, and improv theater. I still consider myself, first and foremost, a writer; when people ask me what I do, that's what I say.

The more I've learned about writing, the more I realized that good story structure is more analytical than most people think. One thing that seems to stymie many beginning writers is structure: dealing with the nuts and bolts of writing. I met a surprising number of writers who had some kind of technical background, and I began to see more and more similarities between science and storytelling.

It was that realization that eventually led to this book. It represents my attempt to fuse the two things about which I am most passionate into one logical whole. I have done my best to avoid advice about any current technological fad or fashion, because I wanted to write a book that would still be useful in five or ten years. To that end, some of my comments are general rather than specific; if it seems like I'm being vague, that's why. My goal is to show you that it is possible to bring a sense of wonder, curiosity, and imagination to the task of science communication. It is possible to retain that sense of awe I felt about those binary stars, while pursuing them with the dilligence of my colleague.

Lastly, I want to welcome you. I wrote this book because I hoped it would help people, but it can only do that if people read it, so thank you.

Let's begin.

Part One

Theory

1. The Challenge We Face

CHORUS

```
Oh for a muse of fire that would ascend
The brightest heaven of invention!
A kindgom for a stage, princes to act,
And monarchs to to behold the swelling scene!
```

-HENRY V (~1599)

Over the course of the last five centuries, the human species has orchestrated a transformation of itself and its world. In scarcely more than twenty generations, we have fundamentally altered ourselves, our relationship to the natural world, and the face of the planet that gave us birth. In the early 16th century, daily life for the vast majority of humans would have been familiar to people from two, or five, or ten thousand years earlier. People in many regions of the world were still hunting and gathering, living in small tribes or clans, while others were practicing the type of simple subsistence farming that their ancestors had been using since prehistoric times. Even in the most sophisticated and advanced European cities, where a Renaissance of classical ideas was gathering momentum, few dared dream that art or engineering would ever be as sophisticated as they had in antiquity. Even as the Age of Exploration approached its zenith, and a handful of Europeans discovered (and quickly began to exploit) previously unknown lands on the far side of the world, the vast majority of people rarely travelled more than a few miles from home. In the early 16th century, most people lived simply, died young, and thought of the world as a very big and very mysterious place.

Five hundred years later, human existence has undergone a complete metamorphosis. Human hands have touched (and in many cases marred) even the most remote corners of the Earth. Humans are connected to one another in ways that seemed impossible only a century ago. We think nothing of travelling across a continent or an ocean in the course of a single day, and we chat amiably with people on the other side of the globe without a second thought. Most of us live very long and healthy lives, protected from the capricious fortunes of famine or disease. In the industrialized nations, we are almost completely insulated from the most basic aspects of life; the production of food, the collection of water, the construction of shelter. In just five hundred years, human beings have shrunk and homogenized the world, and managed to build a global society that enjoys levels of peace, prosperity, and luxury that our ancestors would have been unable to even imagine. And nearly all of that change is the result of science and technology.

"Science," has at least two distinct and very closely related meanings in modern usage. The first is what we who practice it usually mean when we use the term: an interconnected global enterprise that systematically builds and organizes knowledge about observable phenomena, using observation and experimentation to build and test hypotheses. Yet the word has also come to mean the body of knowledge which that process generates and shares. The word "Science" first came into common usage sometime during the early nineteenth century, as huge advances in chemistry, biology, botany, geology, and a host of other fields made it necessary to coin a term which would encompass all these pursuits. At the same time, that portion of what was then called "Natural Philosophy", and is now known as physics and cosmology, was separated from its roots in pure philosophy and joined with the rest of the sciences. It was during this period that truly modern science was born, science (Latin, meaning *knowledge*), entirely divorced from philosophy (Greek, meaning *Love of Wisdom*).

Since that separation, the scope of both the practice of science and the knowledge generated thereby have exploded. Our ability not just to understand and describe the world around us, but to manipulate that world, has given the human race powers that the ancients believed were reserved for the Gods. Modern medicine, modern transportation, modern communication and modern agriculture would all seem to a visitor from ancient Athens like blessings bestowed by the denizens of Olympus. Science and technology are now so closely woven into the fabric of our existence that we can no longer escape them; even the most rugged journey to the remotest part of the globe is tracked by satellite, and even the poorest farmers in the world's least developed nations know how to use a smartphone. Science and technology have changed every aspect of the way we live and interact with one another, and have altered the face of the planet to such a degree that a visitor arriving from a distant star would be able to tell with a single glance from a great distance away that ours was a densely populated and technologically advanced world.

Yet while science has given us enormous power, it tells us nothing about how to use it. The strength of science is that it offers a purified and crystallized process for rational thought, a process that has been purged of corrupting influences like spite, or jealousy, or pity, or compassion, or love. Science's divorce from philosophy was not merely a semantic change, but a conceptual one; science cares only for knowledge, it eschews any pretense of wisdom. It is for this reason that many in the modern world have come to fear science, and regard those who practice it with suspicion. Many of these fears are based in superstition, or on misinformation, while others are reasoned responses to clearly established facts. For all the comforts and conveniences that science has brought to humanity, science has also enabled the worst of our offenses and atrocities. Climate change, mass extinctions, pollution of the water and the air, eugenics, genocide, and the never-ending drive to build more and more destructive weapons of war must also be laid at the feet of science. For all its power and sophistication, science is a tool like any other, and tools are neither intrinsically good nor intrinsically evil, they merely magnify the qualities of the people who wield them.

Meanwhile, as science has become more and more sophisticated, the complexity of the language used to describe it, and the level of background training that is required before one can meaningfully participate in it have grown exponentially. We expect students who wish to pursue careers in the sciences to have completed nearly a decade of background instruction before they truly begin their formal training in a particular field. Some disciplines have become so diverse and so specialized that two scientists in the same department may have little or no working knowledge of the other's line of inquiry. Technology too has become so complex that we routinely build devices whose complete inner workings are fully understood by no one person. Indeed, technology has become so seamlessly integrated into modern life that many of us have come to view it as a natural consequence of existence, as part of the world around us, and lack even any interest in understanding how it might work. Taken together, these trends mean that most humans know less and less about an increasingly powerful field that they depend upon more and more. Or, to appropriate a famous line of Winston Churchill's: "Never was so little known by so many about so much."[1]

For all of these reasons, clear, powerful, and effective communication of scientific ideas has never been more important than it is today. Science is a huge, diverse, interconnected global enterprise that undergirds most of the infrastructure of the modern world. Yet that enterprise depends critically on the free and open flow of information and ideas, both between and among disciplines within science, and between scientists and the those who use it,

1 The original is: "Never was so much owed by so many to so few," praise given by Britain's Prime Minister to the pilots of the RAF during their defense of the island from German warplanes in the summer of 1940.

fund it, rail against it, or otherwise feel its impact. Communication is not a thing that we do once the science is complete. Communicating science is an integral part of being a scientist.

What This Book Is

This is a book about communicating science, designed for people working in the sciences, or in the many data-driven fields that are science's closest relatives, including engineering, medicine, the social sciences, etc. (henceforth together referred to as "scientists"). A considerable portion of the arguments presented herein will address challenges or situations commonly faced by academic scientists, but should be equally applicable to those working in industry, government, or other sectors. Every effort has been made to keep what appears in these pages firmly rooted in the practical and the applicable, rather than straying into the conceptual or theoretical. An enormous literature exists on the communication of science, which has thrived as a social science in its own right for more than three decades. Yet much of that literature focuses on what might properly be called macro-communication; examining trends or looking for patterns on a regional or national scale over a period of years or decades, or looking for systemic or institutional explanations for particular successes or failures in the communication of particular topics. This literature, while hugely informative, rarely has concrete advice for the individual scientist who faces the problem of sharing a particular technology or research result. This book aims to fill that gap, by providing clear and easily implemented advice about how to approach several of the most common questions in scientific communication, in hopes that the reader can begin to improve his or her communication technique.

What This Book Is Not

This book is not a work of scholarship. It is not built upon the foundation provided by any existing literature in any field of study, and as such does not cite sources.[2] The information herein derives from a great many sources, but is largely based on the experiences of the author in training scientists to communicate more effectively. The goal here is not to advance the study of science communication, but rather to provide scientists with a working guidebook on how to best communicate, presented in the most concise and readable format possible.

This is also not a textbook, or at least is not primarily intended to serve that role. While there may be some instructors who choose to include this tome as part of the required reading for a course about communicating science (and it would be a great benefit to science and to society if more such courses existed and were regularly offered), this book's chief audience is not students of science or of science communication. These pages are written for working scientists, those of us at or near the end of our educations, who are engaged in scientific or technical work, need to share that work, and want some practical advice about how to go about it.

It should also be noted that this book does not address the particular skills required to write a scientific paper, nor does it purport to tackle the subject of data visualization. Publication has become so central to the career of a scientist that considerable attention has been paid to the question of paper-writing, and several excellent texts on the subject are available. What's more, specialized forms of the scientific paper, with their own unique conventions, have emerged in many fields, and little can be said that is applicable across disciplines. Likewise, the techniques used to visualize data vary widely across fields. Because this book aims to provide instruction in

2 A number of books and other materials are noted for reference or for further reading; references will appear in these footnotes throughout the book.

areas not covered in most formal curricula, and to address the challenges of scientific communication common to all of science, these two subjects will be left aside.

Lastly, this book is not about teaching science. Much of the advice in this book is absolutely applicable to presenting material in the classroom, and it is true that good communication is one of the backbones of good teaching, but excellence in teaching requires far more than effective communication. (Indeed, at times, the skilled teacher is trying to conceal or obfuscate information, rather than clarify or expose it). It remains the case that many working scientists are asked to teach science to young and aspiring scientists without any training whatsoever in proper teaching technique. This deficiency is lamentable, but not something that this book purports to address.

The Role of Communication

Many scientists (and equally many non-scientists) view communication as a corollary or supplemental activity to scientific research. In their view, the process of scientific investigation or technological invention is complete once results have been obtained and verified or a product built and tested. Yet as we noted above, this view ignores the fact that the scientific enterprise is fundamentally dependent on the ability of researchers who may be separated by great spans of time or distance to share information with one another and with the rest of the scientific community. Even the most compelling or insightful research, the most groundbreaking theory or discovery, only truly becomes science (either the search for knowledge or the knowledge itself) once it is shared. Communication is the cornerstone of collaboration, and collaboration is one of the cornerstones of science. Far from being an ancillary or supplemental activity, communicating the results is part of doing the science.

Despite its importance, the current process for the communication of scientific results leaves much to be desired. A huge percentage of scientific findings are still reported primarily through the academic publishing system, a model built on paper and printing that has changed little since its development in the late nineteenth century. As digital technology becomes ubiquitous, the institutional apparatus for sharing scientific results is beginning to buckle under the stress. New models, including open access and online journals, are becoming more and more common, and some of the institutions that fund scientific research are beginning to demand that the results generated by those funds be made publicly available. The model for sharing scientific findings is beginning to evolve, and though we cannot predict what form the new model might take, the forward-thinking scientist would be wise to follow that evolution closely, and add their voice to the conversation about how it should proceed.

Beyond the role of communicating science to the scientific community, scientists also have a growing need to share the results of their science with the wider world. As the number of scientists around the world continues to increase, and the number of questions that science can explore continues to multiply, the competition for resources, in terms of funding, facilities, and personnel grows ever more fierce. In that competition, the scientist who can communicate most effectively about what her work is and why it is important will have the upper hand. What's more, as many of the most pressing issues faced by societies around the world revolve around questions of science or technology, scientists themselves are asked to take a more visible role in communicating science to specific publics.[3] Finally, the ability to communicate in clear, engaging, effective language, and to stimulate interest in and understanding of one's work is one of the biggest differentiators between scientists, and has a huge impact on the likely trajectory

3 This is not a typographical error. Social science research has demonstrated again and again that the concept of a "general public," is fallacy; rather there are several possible publics, groups of people who coalesce around something they all have in common.

of a scientist's career. Defending a doctoral dissertation, getting hired into a lab, recruiting students or employees, securing funding or budgetary approval, publicizing results, supervising employees, and interfacing with superiors all depend critically on one's ability to communicate.

Unfortunately, this critical skill is rarely included as part of a scientist's formal education. There are many possible explanations for this deficit, but the one cited most frequently is the simple argument of scheduling. Science has become so vast and interconnected a body of knowledge that becoming proficient in even its most basic tenets requires years of diligent study. The student who aspires to become a scientist must tackle a host of complex subjects, must become proficient in advanced mathematics, and must master the sophisticated technological tools that are almost always now at the core of any line of inquiry or experimentation. As a result, most scientific educations leave little room in the schedule for other pursuits. We scientists like to think of ourselves as well-educated people, but in truth, though our educations are very deep, they are seldom very broad. While we may learn a great deal about the functions of various cells, or the movements of the stars and planets, or the structure and properties of various materials, we often know very little about poetry, or music, or literature, or sport. Yet these things are also part of the human experience, and each in their own way can motivate and inspire change.

The result of this lack of any formal training in communication is that most scientists end up learning communication on the fly, mostly by simply imitating the style of communication they see employed by their professors, advisers, and mentors, and that style is often bad. Because the most senior generation of scientists was similarly denied any training in communication (indeed, a generation ago, communication, especially public communication, was something many in the scientific community viewed as a wasteful and unnecessary distraction), and because during the course of their careers the technology of communication underwent a significant shift, an ad hoc style emerged. Despite the many shortcomings of this style, it was quickly imitated and began to spread. One of the goals of this book is to dispel some of those common misconceptions about what constitutes an effective style of communication, in the hope that this trend can be reversed.

To that end, we close this chapter by challenging a frequently asserted misconception, and establishing one of the fundamental tenets of good communication that we will return to in later chapters. In much of the available literature about scientific communication, many authors draw stark distinctions between what is called "professional communication," and "public communication." Publics, so this argument goes, have different needs, speak a different language, and expect a different style than do an audience of scientists. While you may be very good at speaking at a scientific conference, in other words, you do not possess the necessary skills to speak to a more general audience.

This is a false dichotomy and a dangerous fallacy. There is no line of demarcation between communicating to a group of peers and communicating to any other audience. While it is true that you might employ slightly different techniques when defending a doctoral dissertation than when speaking to a group of schoolchildren, the core principles of effective communication are identical in both cases. While every audience is different, and each piece of communication should be tailored to fit the needs of that audience, every audience is also the same, and there is a short list of things that every audience wants.

Artificially dividing public from professional communication is dangerous for two reasons. First, it leads scientists to believe that the two are separate skills, that the principles and practices of successfully communicating to the public are things that must be learned separately from and in addition to the principles of communicating with their peers. Worse, however, is the fact that this false dichotomy leads many scientists into the mistaken belief that even though they struggle to convey their ideas in general terms, they are quite capable of successfully sharing those ideas to a technical audience. This is simply not the case. The fact is that the most important principles of good communication are universal, and are independent of audience. A scientist who general audiences find dry, stiff, boring, pedantic, or unintelligible will be every bit as dry, stiff, boring, pedantic and unintelligible when speaking to his peers. Conversely, a scientist who can convey her work in a crisp, dynamic, engaging and intelligent fashion, using language the audience can understand, will have success regardless of whether her audience is senior colleagues from her field or a room full of people who know little or nothing about science. Good communication is good communication, and one of its hallmarks is careful consideration of and attention to the listener.

2. Not A Science

```
                HAMLET
There are more things in heaven and earth,
     Horatio,
Than are dreamt of in your philosophy.

                              -HAMLET (~1601)
```

Science is one of humanity's most powerful tools for solving problems, and people who are scientifically trained tend to apply scientific thinking and a scientific methodology to problems of all types. Communication is no exception. Many scientists view communication as essentially an information transfer problem, one that can be profitably studied and analyzed using the language of information processing. We envision information pathways, through which signals are transmitted and received, we think of bandwidth, and we struggle to minimize noise. Yet while information transfer does fall under the definition of communication in the broadest sense, it is not the subject of this book. Here, we are concerned not with the passage of data from one node to another, but in the transfer of ideas, thoughts, and feelings between human beings. While some of the language of signal processing is used in these pages to express a point, its use is largely metaphorical, and is not meant to imply that one can truly understand or practice effective scientific communication merely by tackling the problem scientifically. The process by which people share information is necessarily more complex, more chaotic, and more dynamic than the simple transfer of data, because it involves human intelligence and human emotion. Emotions are non-deterministic and non-repeatable, they are mutable, and they are evanescent. The transfer of emotion from one human being to another is not something that can ever be fully understood through the sciences, which is why it has always been the central focus of the arts.

Apollo and Dionysius

The role of emotion in the process of communication is sometimes controversial, and something that many scientists approach with hesitation. The notion of making an emotional appeal, or trying to connect with an audience on an emotional level, seems anathema to most scientists, because we have been trained to view emotion as a corrupting influence; a force which can distort or pervert good science. The great strength of science is that it can generate results that are independent of any personal belief or conviction; experiments and observations are only verified and considered valid when an independent and impartial observer obtains the same result. The success of science depends on the ability to conduct research without any preconceived notions or emotional attachments to the subject under study. The major scientific revolutions of the last several centuries were all dependent on this detachment, and in most cases were triumphs of hard facts that stood in stark defiance to deeply rooted doctrine or dogma.[1] Good science requires a commitment to logic and reason that exist independent of emotion, and scientists often fear being perceived as emotional, thinking it will compromise their integrity as scientists. Yet the scientific mind, the rational mind, the portion of the human psyche that is methodical and analytical, represents only one section in the orchestra that is human

1 The best examples of this are the Copernican Revolution in cosmology, and the Darwinian Revolution in biology. The interested reader should consult Thomas Kuhn's *The Copernican Revolution*, (1957), a masterful work of history that contains many of the foundational ideas that would undergird Kuhn's later and better-known work.

consciousness. Ours is not a species that operates on logic alone. We are also sensual, creative, and chaotic. Human beings rely on both intellect and emotion.

The notion of this dichotomy between intellect and emotion has existed in philosophy for millennia. In popular culture, the divide is often referred to as "left-brain" and "right-brain" thinking, despite the fact that neuroscience has consistently debunked the notion that the two forces dwell in distinct hemispheres of the brain. The language of philosophy is more instructive, labelling these two impulses the Apollonian (after Apollo, the Greek God of the sun, God of truth and reason) and the Dionysian (after Dionysus, God of wine and ecstasy). The dichotomy is sometimes also expressed more colloquially as the distinction between "head" and "heart," the part of us that thinks, and the part of us that feels. This notion has persisted for centuries because it rings true for most people; we feel within us a separation between our passionate impulses and our restrained calculations. While these two forces are sometimes in opposition, the process of human decision-making relies on bringing them into alignment. The Apollonian mind can convince of us truth, but the Dionysian mind can compel us to action.

For much of human history, the two forces were viewed as necessary complements to one another. It is neither an accident nor a coincidence that many of the greatest thinkers from antiquity through the Renaissance, from Aristotle to Leonardo, were true polymaths, thinkers whose lines of inquiry spanned the full breadth of the human experience, and whose works in the arts were every bit as ground-breaking and influential as their accomplishments in the sciences. Only in the last one or two centuries have we bifurcated the two minds, and begun to think of the arts and the sciences as two distinct and independent fields of inquiry, often to the detriment of both. The two disciplines are more interconnected than many appreciate; music and sculpture are underpinned by mathematics, and cosmology and paleontology require great leaps of creativity.

Almost all of the fields of human endeavor that venture into the realm of human persuasion firmly embrace both sides of the psyche; they appeal both to head and to heart. In politics, or advertising, or entertainment, or the arts, or the law, one consistently finds both intellectual and emotional content. It is when the two impulses work in tandem that they have their greatest impact, for there exists some interaction between them that allows their combined whole to exceed the sum of their individual parts.

Most scientists think of ourselves as rational creatures, as people who are firmly rooted in the Apollonian mind, at least in our professional lives. And so when asked to share the results of our work, we tend to stay firmly rooted in that realm, because we feel our credibility as professionals depends upon it. Yet we are also emotional creatures, who have broad and varied interests in many other fields beyond our professional careers. Some of us are musicians, or lovers of literature, some are intrigued by the culinary arts, or enthralled by the theater, some of us are passionate about athletics, and some of us are deeply devoted to our religions. Nearly all of us love our families, and are committed to our communities. Though some of us may struggle to express ourselves, we feel deep emotional connections to many things. Indeed, a great majority feel a deep and abiding passion for the work that we do, but we fear allowing that passion to show, lest the quality of the work be called into question.

One of the central theses of this book is that emotional content and emotional connection with an audience is one the pillars of effective communication in any field, including the sciences. As we shall see in later chapters, emotional connection does not require any special magic, nor does it require the communicator to expose him or herself in an uncomfortable way. All that is required is to identify or create something that the audience cares about, and link the content of the communication to that thing.

As noted above, most scientists are hesitant to embrace this approach, and fearful that it may compromise their reputations or the perception of their scientific integrity. Yet (as we shall see) it is perfectly possible to do good science, to conduct observation or experimentation in a completely dispassionate way, and still find a way to let the communication of that work connect to an audience emotionally. Most of the remainder of this book is dedicated to showing you how that is possible, and it has never been as important as it today.

"Facts, facts, facts!"[2]

For the vast majority of the last several hundred years, the primary goal of scientific and technical communication was to preserve and disseminate factual information. Science has become one of humanity's greatest intellectual achievements precisely because it was successful in this endeavor. The careful preservation of data and the wide circulation of new theories meant that scholars thousands of miles or hundreds of years distant from one another could collaborate in a meaningful way. Scholars through the eighteenth century were well aware that knowledge could be lost as easily as it could be obtained, and our current system of academic literature evolved at least in part to make sure that our hard-won knowledge would never slip through our grasp.

At the same time, the ever-increasing pace at which scientific discoveries were made continued to have a huge impact on the daily lives of most people. New cosmologies in the seventeenth century literally re-made the world, and new technologies in the twentieth nearly destroyed it. For nearly four hundred years, communicating science was largely a question of communicating the facts, because the facts were hard to obtain and easy to lose. When we sought to share science, we sought to share the facts, because

2 Thomas Gradgrind, speaking in Charles Dickens' *Hard Times* (1854), a novel which lampoons education in Victorian England.

the facts were all that mattered.

This once-sacred truth has vastly changed in the modern age, as a result (once again) of technology. Facts are no longer sacred. Facts are cheap. We live in an ever-expanding sea of facts, a sea so vast and broad that many of us feel submerged in them; like the Sorcerer's Apprentice, the trickle we fought so hard to conjure has now become a deluge. Information has never been as plentiful or as easy to obtain as it is today. Indeed, entirely new fields of science are now evolving whose subject is data itself, finding ways to extract new (or useful) information from the torrent of data we daily generate.

To the average citizen, this proliferation of information, and the effortlessness with which it can be obtained, has eroded our sense of wonder. Information once greeted with gasps of incredulity is now met with shrugs of indifference.[3] The answer to any factual question is rarely more than a few keystrokes away, and despite the criticism often levied against them, the quality of those answers is usually quite good. The cost of information is falling asymptotically toward zero, and as it does, the cost of enthusiasm rises asymptotically toward infinity.

This proliferation of data means that it has never been easier to find information you are looking for, but it has never been harder to find information you are *not* looking for. The ascendancy of the search engine has been the death of browsing. One is far less likely to simply stumble across an article or book these days, because there are more of them at our fingertips than ever before. One can find almost any piece of information one wants, but only if one has the motivation to seek it.

In this new world, it is more important than ever for scientists to be able to connect with an audience in a way that generates interest, enthusiasm, or excitement. We can no longer be content merely to communicate the

3 Those who would dispute this assertion should spend more time with the teenage population of any developed nation.

facts, because the facts can be packaged and transmitted to anyone and everyone across the globe in a matter of microseconds. Instead we must embrace the task of communicating what those facts mean, and why those facts matter. We, as scientists and communicators, must embrace both the Apollonian and the Dionysian psyche, must engage both head and heart, if we want our message to be a signal which stands out above the noise.

Now more than ever, communication, even scientific communication, must fundamentally connect with an audience's emotions. And by definition, such communication is not a science, but an art.

3. Order of Operations

JOE
Audiences don't know somebody sits down and writes a picture. They think the actors make it up as they go along.

-SUNSET BOULEVARD (1950)

Scientists, despite enjoying widespread respect throughout society, have a nearly universal reputation for being poor communicators, and it is not altogether undeserved. While there are many among us who are excellent in front of an audience, capable of rousing the interest and piquing the curiosity of listeners regardless of their background or level of technical expertise, there are equally many who are apt to drone endlessly on, to fill a blackboard or page with a deluge of technical information that is far beyond the abilities of the listener or reader, and muddle incoherently through a series of facts without ever reaching any palpable conclusion. While it is not fair to say that most scientists are poor communicators, it probably is accurate to say that many scientists communicate poorly. There are many possible explanations for this deficiency, but from a practical standpoint they can be essentially distilled to three main issues, all of which are derivations of a type of cognitive bias known as the curse of knowledge.

A term which first emerged in the language of economics, the curse of knowledge refers to the phenomenon that well-informed people, with deep knowledge of a subject, have a very difficult time viewing that subject from the perspective of someone less well-informed. This bias has been shown in a number of experiments, the best known of which involves a subject tapping out the rhythm of a well-known song to an audience. The person doing the tapping usually estimates that something more than 50% of the audience will be able to guess the song from the rhythm alone. This is because while tapping out the rhythm, the subject is also hearing the melody (and probably also the accompaniment) in his or her head. The audience, meanwhile, with only the rhythm as a clue, lacks that knowledge, and typically less than 5% of an average audience can name the song.

This curse is felt particularly acutely by those of us in scientific and technical fields, because of the depth of knowledge that we possess. As discussed in earlier chapters, a scientific education is of necessity a deep one, in which several layers of increasingly complex schema are nested within one another like Russian dolls, and we must master each one of them in sequence to arrive at a true and deep understanding of a particular phenomenon. What's more, the list of possible phenomena to investigate is so long (and constantly growing), that those of us in a very specialized area may be one of only a tiny handful of people in the world who truly understand the thing we are trying to explain. Exacerbating this problem is the fact that many of us spend the majority of our time talking to and working with people who are involved in the same or very similar work, and so we become comfortable using shorthand language that may be unintelligible to those outside our discipline, and accustomed to launching into very complex technical discussions of our work with little or no preamble. One of the reasons that many scientists have trouble communicating effectively is that most scientists speak most frequently about the thing about which they know the most. As a result, they are almost always talking to people who know less about their subject, and this can lead to problems, because the speaker assumes that the audience knows too much.

The second problem in the sciences that seems to exacerbate bad practice in communication is what we might call the curse of curiosity. Scientists spend most of their time trying to explain things to other scientists, and scientists are curious people, who quite like things that are difficult to understand. We like puzzles, and riddles, we are drawn to mazes, and we especially like learning new things about science. The reason this is a problem is that it makes us far too tolerant as audience members of communication that is sub-par. When confronted with a presentation or a poster or a tutorial which confounds us, we tend to stop, and pore over it more closely, straining to tease out the underlying meaning. And if we fail, we almost always blame ourselves, our own lack of knowledge or lack of comprehension, rather than placing the blame for that failure on the communication itself. Because we are curious, we lift the burden of good communication off of the communicator, and place the burden of comprehension on ourselves, which is the opposite of what we should do.

The third, and perhaps most insidious problem that seems to affect science communication is what one could call the curse of fame. As we eluded to earlier, the curse of fame is an extension of the lamentable fact that communication is simply not valued as a professional skill in the sciences, or at least hasn't been valued historically. We still judge the eminence of a scientist solely by the quality of his or her research work, with little or no weight given to how well he or she is able to make him or herself understood. What's worse, a scientist can rise to a considerably senior position in many institutions and organizations without ever gaining any experience as a communicator, especially outside of his or her discipline. Yet as audience members, we are eager to hear findings and conclusions presented by the scientist who is their author. Thus a scientist who has garnered fame for her research will be asked to speak in a great many places to a great many people, even if she is a notoriously bad speaker, and the audience will likely sit in rapt attention, patient and curious, but understanding little and remembering less.

The curse of fame has been a pervasive problem in scientific communication for a generation or more, and it is one of the reasons why you find so many senior scientists who are not only poor communicators, but believe themselves to be quite skilled. Because their audiences always applaud, and because the speaking invitations keep coming, they come to believe that they have mastered a skill which they are in fact bungling. These scientists are thus often resistant to the notion that they can or should improve, and (far worse), end up with a cohort of students or junior colleagues who mimic their speaking style in hopes of someday recapitulating their scientific success.

The best way to avoid the curses of knowledge, curiosity, and fame, is to approach each piece of scientific communication using a specific order of three distinct operations. In this section we will refer to these steps as they are implemented when preparing for a public talk, but precisely the same operations should be used regardless of the medium or platform through which you communicate. The three steps are: choosing your impact, selecting your audience, and crafting your message. Or more succinctly: choose why, choose who, and choose what.

Choose Your Impact

When preparing to communicate, particularly to a general or non-technical audience, most of us quite logically begin the process by asking ourselves: "What should I say?", and this is absolutely the wrong question to ask at the outset. The reason it is the wrong question stems from the curse of knowledge introduced above. You know a great deal, perhaps more than anyone else in the world, about the subject of your particular expertise. The list of things you *could* say about that work is very, very long, and stretches from the earliest historical underpinnings of your work all the way to potential applications or implications that could be decades or centuries in the future. The number of potential stories you could tell about that work is practically infinite. Many scientist fall frequently into this trap, simply picking out the first and obvious thing they could say about their work. Yet these communications often fail, for this haphazard approach to selecting your content tells you nothing about what you *should* say. The first step in your preparation, then must not be: "What should I say?", but rather: "Why am I speaking?"

We are often tempted to answer this question flippantly, with some variation of: "Because they asked me to speak," but the question really does bear careful consideration. Why are you speaking? What is it that you hope to achieve? It is often helpful to think of this question in mathematical terms: you are trying to affect some change of state on some variable. What is it? Typically, this means changing what some person or group of people knows, thinks, or feels. This change can be very simple, or it can be quite complex, it can range from the trivial to the profound, but you the speaker should have the change firmly in mind. We call this step choosing the impact, and it is the final goal, the desired result, of your communication. Before you write a single word, or select a single figure, before you even begin crafting what you want to say, you should be able to succinctly state your intended impact.

Developing a good impact statement is a critical first step, for two main reasons. First, it acts as an effective first-pass filter, greatly reducing the number of possible things you *could* say. Second, (and perhaps more importantly) it gives you a metric by which to judge the success or failure of your communication, and thus a means of comparison by which to judge subsequent changes in your style or content. To these ends, a well-crafted impact will have three main attributes: it will be specific, it will be achievable, and it will be (ideally) measurable.

A specific impact is one that truly focuses on a specific change. You want the audience to leave the hall knowing, thinking, or feeling something that they did not know, think, or feel before your talk. Articulate what that is. It is not enough to say: "I want the talk to go well," or "I want the audience to applaud at the end." Those are impacts about *you*, not about the audience. Choose what change you want to make on *them*.

Similarly, an achievable impact is one that can be affected by the audience you have. If preparing for a talk to a local primary school, the passage of a landmark piece of legislation cannot be your chosen impact.[1] Likewise, a measurable impact is one that can be observed. While it is not always possible to measure this with any degree of precision, it is often possible to solicit individual responses, which can help give you a general sense of how well you succeeded. The ability to succinctly state a desired impact which is specific, achievable, and measurable is the critical first step in constructing any piece of communication.

Identify Your Audience

The second step in crafting a successful piece of communication is identifying the audience for your communication. Many of us seem to skip this step, because we tend to assume that the audience is dictated for us by the format of the communication itself. While this may at times be true, it is almost always the case that audiences are heterogenous, in terms of their level of scientific literacy, their level of interest, their pre-conceived ideas, etc. What's more, it is a mistake to assume that your audience necessarily comprises the whole population of people who are reading, watching, or listening to what you have to say. You may in fact be targeting your message to a very specific (and possibly quite small) subset of that population. Luckily, choosing your specific audience is easy, and is largely a *fait accompli*, if you have carefully crafted your desired impact. Your audience is the target for your impact.

1 In this circumstance your desired impact might be that the audience leave the hall believing that the legislation should be passed, an important distinction.

While it is true our audiences are in some circumstances dictated to us (a presentation to the Board of Directors of your company, say, or a defense of your Ph.D. Thesis to your advisory committee), the selection of impact and audience must go hand in hand. If commanded to speak, decide what impact you wish to have on that audience. If, however, you are invited to speak, decide what your desired impact will be before you accept. If you cannot formulate an impact for that audience, it may be wise to decline the invitation. Once you have both your audience and your desired impact in hand, you must identify two critical attributes about your audience: their level of knowledge, and their level of interest.

Determining the level of knowledge of your selected audience is absolutely essential, and it is a place where many scientists go wrong. As a result of the curse of knowledge, we almost always tend to pitch our content at too high a level of technical detail for our audiences. This is true not only in general public communication but in professional communication as well. Even at a symposium during a technical conference, it is not uncommon to see a scientist giving a talk that largely flies over the heads of the audience, yet the audience either blames themselves (curse of curiosity), or is too polite to say anything (curse of fame). One of the reasons we tend toward the overly technical is that because the concepts are so familiar to us, we fear boring the audience with what seem elementary explanations. This fear is almost entirely unfounded. Audiences are rarely bored by hearing someone explain something they already understand; indeed it often fills them with a sense of pride, particularly if the explanation is good.

This overestimation of the appropriate level of technical detail is especially important in the twenty-first century, because (as we discussed in Chapter Two) it is trivial for an interested audience member to obtain more technical information, yet nearly impossible for her to extrapolate the larger narrative on her own. When preparing to speak to an audience for the first time, it is helpful to do some background research about what kind of people will likely be in that audience. Look for footage of others who have spoken to the same audience. Ask your hosts for their feedback. Do some research on who is likely to attend the event, including age, level of education, etc. One common mistake to avoid is the presumption that well-educated people will have some background in science. There are many very educated, very intelligent people who know nothing about science, just as there are many intelligent and well-educated scientists who know very little about much else.

There are often circumstances in which you will be asked or invited to speak in a venue or medium in which it is difficult or impossible to predict the level of knowledge of the audience.[2] In these cases, it always best to err on the side of the general over the technical. The reasons are twofold. First, as outlined above, a general talk will not alienate a technical audience, while a technical talk will alienate a general audience. Second, (and perhaps more importantly) it is relatively straightforward to add technical detail to a general talk (e.g. in response to a sophisticated question), while it is nearly impossible to extrapolate general principles from a talk which is rife with technical details.

Once you have gauged or made an educated guess about your audience's level of knowledge, you must next make an effort to determine their level of interest. Luckily, this is often easier than determining their level of knowledge, for most people have an interest in scientific topics that exceeds their understanding. Developing a sense of the audience's level of interest is something that should take you back to your impact statement, because understanding what they currently know, think, or feel is critical to understanding how what they know, think, or feel can change. Once again, the way to go about this to do some research; ask your hosts, look for a record of previous similar talks, make an educated guess. In some cases, this job

2 A lecture at a science museum, or an appearance on a radio call-in show, for example.

will be easy, as audiences for scientific communication sometimes develop and congregate around particular issues, and their level of engagement is very high.[3]

If you find it difficult or impossible to gauge the interest of the audience, or if you simply have no information whatsoever, it may help to remember the simple rule that people tend to be very interested in things that affect them directly, and much less interested in everything else. It is for this reason that the more applied your work is, the more engaged an audience you are likely to face. An oncologist researching a breast cancer treatment that is about to start clinical trials is far more likely to find a deeply engaged audience than a mathematician working with a new orthogonal family.[4] Where possible, try to link your work to possible outcomes or applications. If that is not possible, use historical examples to show how previous similar work may have enabled an existing application. The less familiar the audience is with your field, the more important it is that you find a way to connect that work to the world of their experience.

Craft Your Message

Once we have these critical pieces of information in hand, then and only then can we begin to craft our message. As we said, we cannot ask the question, "What should I say?" until we have specific answers to the questions: "Why am I speaking?" and "Who am I speaking to?" You will undoubtedly find that in the process of answering these questions, you have necessarily filtered out a huge number of possible choices of what you could say, and are now dealing with a far more manageable process when selecting what you should say. Indeed, if you have completed the first two steps carefully, the answer tumbles neatly out, the things you should say

3 More on this can be found in Chapter 12, Controversy.

4 Pure research has long been, and will likely continue to be, a hard thing to sell.

are the things that will most effectively drive your chosen audience toward your chosen impact. This will automatically dictate several aspects of how you communicate. In choosing the appropriate level of language, you must choose the language that your chosen audience will best understand. In choosing your line of reasoning or argument, you tread a path that moves toward your intended impact. And in choosing what evidence or results to show, you should choose that information that most clearly supports your chosen line of reasoning. Notice that this is something that would have been infinitely harder if we had begun by simply choosing our content. We would be forced to evaluate each potential storyline for an impact and an audience, the equivalent of hunting and pecking for a solution. By following the order of operations, we are able to isolate the variables one at a time, and develop a more analytical solution which leads us more directly to the appropriate content.

Once we have that content in hand then, we are ready to begin structuring the communication itself. We have chosen our pieces, but must now choose their order of assembly. And in this, we will be greatly aided by employing what is perhaps the oldest art form known to our species: narrative storytelling.

4. The Most Ancient Art

Tell me of the storm-tossed man, O Muse,
Who wandered long after he sacked the
sacred citadel of Troy.

-THE ODYSSEY (~800 B.C.)

Among anthropologists, there is broad agreement that *homo sapiens* of 50,000 years ago were genetically nearly identical to modern humans. Though the advantages of health, hygiene, nutrition, and technology have made modern humans physically different in many substantial ways from those distant ancestors, were it possible to implant an embryo conceived by a couple in the late pleistocene into the womb of a modern woman, we have reason to believe that a resulting boy would grow into a man indistinguishable from the rest of his generation.

Despite the close kinship we have with those early ancestors, we know relatively little about how they lived. We know that they used fire, and that they made and used primitive stone tools, which were already massive advantages over every other animal on the planet, but they had none of even the most fundamental technology that we enjoy. They had no cities, no systems of writing, no means of transportation, no medicine, no agriculture, and no domesticated animals. 50,000 years ago people were still very much at the mercy of the weather and the wildlife, and their lives were likely brutal and short. Yet they, like us, and like all humans that lived and died in between, did one thing that all humans everywhere have done: They told stories.

The origin of story as a device for the transmission of knowledge, the role of story as the cornerstone of the development of culture, the function of story as a recapitulation of the basic act of cognition, and the advantage of story in the evolutionary struggle for survival, are subjects which we unfortunately do not have space to fully explore here.[1] Suffice it to say that storytelling is one of the oldest and most important facets of human communication, and that the existence of story is a constant that has been observed across every known human culture, no matter how isolated that culture may have been from its neighbors.[2]

The study of the structure of storytelling (also known as narrative structure), and its uses in communication about scientific and technological ideas, is the main object of this book. We shall spend the majority of this chapter discussing that structure in some detail, and giving examples from fields outside of science. In the following chapter we will discuss the theory of that structure within various disciplines of science and as it applies to the enterprise of science as a whole. The remaining chapters of the book are concerned with practical applications of that theory to particular mediums of communication, and the ramifications of that theory in shaping broader topical discourse.

Story's First Theorem

Story is everywhere. We find it in every human culture on this planet, and at all times in the historical record. Storytelling abounds in

1 The subject could (and does) fill several volumes. An excellent popular treatment of the subject can be found in *The Storytelling Animal*, by Jonathan Gotschall.

2 The best example of this may be the Aboriginal Australians, whose culture evolved in almost complete isolation for tens of thousands of years, but whose myths are remarkably similar to those found in Western Europe and elsewhere.

education, in journalism, in popular entertainments, and in the way we share information with loved ones about the comings and goings of daily life. Stories form the cognitive framework through which we understand and contextualize information, and for many of us, constitute the backbone of what we consider to be our individual identities.

Because of the importance of story, the literature exploring the structure and form of the art is immense. As explained in Chapter One, because this book is geared toward providing practical advice for the working scientist, a detailed history of that scholarship is beyond our scope. Instead, we present a simple structural representation of the method of story construction. This representation is synthesized from many sources, but its form is well-established, and a version of it is taught in nearly every course on creative writing, playwriting, screenwriting, etc. Like many of the foundational principles of science, the fundamental concept of storytelling can be expressed in a statement whose simplicity belies its power.

Nearly every human story follows one simple pattern, whose fundamental form can be expressed in just nine words:

Someone wants something badly, but has trouble getting it.

Simple though it may be, this basic principle can be found in nearly all stories told in various cultures around the world. We find it in religious traditions, in stories of genesis and human origin, in myths and folklore, in fiction and in nonfiction. It appears in all the great classics of literature, and in all of the blockbusters of popular entertainment. How this structure emerged is beyond the scope of these pages, but it is enough to know that its basic structure has been preserved, unchanged, for millennia, and it continues to be taught and used throughout the world, because (for reasons we shall begin to understand), it works.

To clarify the relationship of these structural elements, it is sometimes useful to look at the pattern schematically [see figure].

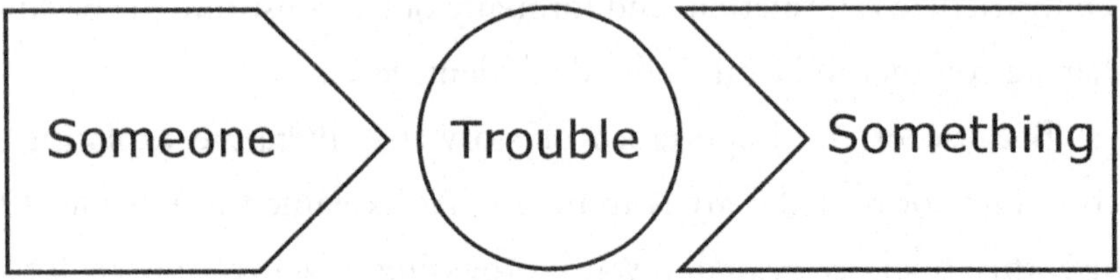

Figure 4-1: The basic schema of narrative. Someone wants something badly, but has trouble getting it.

The structure was first articulated by Aristotle, in his classic treatise on the subject, *Poetics*. The ideas in *Poetics* form the cornerstone of classical literary theory, and it is still taught as the foundational text in many universities to this day. This simple pattern, which Aristotle identified in the great tragedies of the fifth century B.C., continues to define the structure of modern stories, from the cinema to the stage and far beyond. Someone wants something badly, but has trouble getting it.

Because of the long history of this structure and the scholarship that surrounds it, one will often hear this idea expressed using different terms to signify the primary components, but their relationship to one another remains the same. Many modern texts refer to the struggles of a "hero" against a "villain," towards some "goal," where the classical texts refer to a "protagonist" battling an "antagonist" in pursuit of an "objective," but all of these formulations represent the same interaction [see figure].

Figure 4-2: Various terms may be used for the three primary elements, but the structure is unchanged.

The path that the hero or protagonist takes over or through the obstacles in pursuit of the goal is often referred to as the "narrative arc" or "arc of the story" with corresponding increases or decreases in tension or drama as the "rising action" or "falling action", and the resolution of the story is referred to as the "denouement" (from the French, literally meaning "untying"). The particular language used to describe the elements is immaterial, what matters is that they exist. These three simple elements and the relationship between them undergird nearly all human stories, and their use is essential to storytelling success.

Why This Works

Human beings are emotional creatures. We are (despite our frequent protestations to the contrary) largely governed by our emotions. So powerful is the human emotional architecture that it extends outside ourselves, imbuing our species with a unique power of empathy, the ability to experience the emotions of others. When we see another creature in pain, or distress, or in apparent joy, our own emotional system emulates the process we perceive, creating an emotional "echo" of sorts within ourselves. This urge to emulate is so strong that it extends even to inanimate objects; numerous experiments in cognitive science have shown that human beings ascribe emotional attributes (and even impose fully formed story structures) on even the simplest schematic sequences of events.[3]

Of all the emotions human beings are capable, the most powerful, the most primal, and thus the easiest to emulate, is desire. Indeed, desire is so fundamental that it can easily be observed across a broad host of species besides our own. Desire is inextricably linked to our most basic needs for survival: we desire food, and shelter, and sex, and to escape from predators. Indeed, most more complex emotions can expressed as a manifestation of some desire; anticipation is the desire to speed up time, regret is the desire to go back and change the past, envy is the desire to possess what someone else has, and so on. Desire, particularly unfulfilled desire, so permeates our daily life that we are naturally drawn to it when we observe it others. We cannot witness someone wanting something without instantly wanting to know whether or not they are going to get it.

This is why story works, why it is the most powerful and important tool in the arsenal of human communication. When someone wants something badly, the audience that witnesses the wanting makes an emotional investment in the outcome. Suddenly the trouble that someone encounters in pursuit of the something gains an importance, an urgency, that it didn't have before. We the audience want badly to know whether or not the hero will achieve the goal, simply because we know the hero wants the goal, and that is the basis of all successful storytelling.

Stories come in many forms, which vary tremendously in length, in tone, in genre, and which are tailored to suit a host of different audiences. To a casual observer, the outward differences in these forms make them appear to be entirely different modes of communication; it may not be readily apparent what an anecdote told by a friend at a cocktail party may have in common with a children's animated film or the epic novels of Russian literature. Yet all are stories, because all follow the essential formula of

---

3 The most famous of these is an animated film from 1944 made by psychologists Fritz Heider and Marianne Simmel. A search for "Heider & Simmel" on any video sharing platform will likely yield the original footage.

someone wanting something badly but having trouble getting it.

One sees immediately how this simple structure must have existed even in the stories told by our ancestors fifty millennia ago. For the record clearly shows that those ancestors' earliest forms of art were pictorial representations of animals that were important sources of food and raw materials to the people who hunted them. With only those two facts, we see the basic form of a narrative structure emerging. Those ancient humans were someone, and the comfort and prosperity the animals represented was something they wanted. They no doubt often had trouble getting it, and through the story of that struggle, knowledge gained by one generation was passed down to the next.

Stories in Practice

Simple though the theory of story may be, the struggle to successfully apply it has bedeviled storytellers for millennia. Indeed, adherence to these principles does not necessarily guarantee success, for storytelling (like all forms of art) is nonlinear and non-deterministic. A story told on a Monday morning to an audience may fall flat, where the same story told to the same audience on a Friday night may have hit home. The process of storytelling is chaotic and unpredictable, and any particular telling of any story has a chance of failing. Instead, storytelling must be approached as a statistical process, in which one structures the story in order to optimize the probability of success, so that averaged over a large number of listeners or a large number of tellings, the overall rate of success will be maximized.

Here we present four principles of applying story theory to the practice of storytelling, with a brief explanation of why each one is important. The careful student of these ideas will start to look for them in the stories we see around us everyday; they can be found in an everything from advertising spots on television to the stump speech given by a local political candidate.

Establish the Objective Early

The primary reason that story structure is so effective is that it causes the audience to empathize with the struggle of the protagonist. In order for that reaction to take place, it must be made clear to the audience who the main character is, and what that character wants, *in the beginning of the story.* You can think of the relationship between the hero and the goal as the "narrative proposition," it is the primary driving force that moves the story forward. A story is not simply a sequence of events, it is the answer to a question: will the hero achieve the goal or not? The audience wants to know the answer, which is why the story captivates them. The reason they keep turning the pages, or not changing the channel, is because they are waiting to hear the resolution to the question, the payoff of the narrative proposition.

This process can be seen most clearly in the structure of modern films. The vast majority of films are divided up into three "acts", or sections. These acts have a (roughly) standardized length, and similar things happen in each of them. The most important of these by far to both filmmakers and audiences is the first act. Comprising roughly twenty to thirty minutes of screen time, the first act establishes who the main character or characters are, introduces us to the world in which the story will take place, clearly outlines what the goal is and why the main character wants it, and begins to foreshadow what kind of trouble he or she will have getting it.[4] This process is referred to as "setting the story in motion," and it is absolutely critical to the success of a story. By the end of the first act, if the audience does not know who the main character is, and what that character wants, the story is very likely to fail.

It is worth noting that despite its importance, the first act does not occupy much of the story's time. A thirty minute first act constitutes

4 A good first act does many more things as well, which is why many screenwriters consider them the most difficult part of a screenplay to write.

only about a quarter of the film's total length. While stories vary widely in overall length, that ratio should be maintained, so that a 1000-page novel should end its first act around page 250, while a 60-second TV spot should have established a hero and a goal in the first 15 seconds. Nothing is more important to a good story than a good first act, because a story is not a sequence of events, but the answer to a question, and the first act tells us what the question is.

Make the Stakes High

Notice that our statement of the theorem of storytelling does not simply say that someone wants something, but that someone wants something *badly.* It is not enough to state that the main character wants something, you must make clear to the audience *why* that character wants that thing, and that they want it badly. The easiest and most common way to do this is by introducing what is referred to as "the stakes," which can be thought of as the cost of failure. We know that our main character wants something, but what will happen if he or she does not get it? The purpose of the stakes is to heighten the emotional tension felt by characters (and thus by the audience) about the narrative proposition.

The most obvious examples of this can be found in classic quest stories. A hero embarks on a journey to find and retrieve (or in some cases, destroy), an object or item that is invested with great power. Because the audience does not necessarily have a reason to care about that particular object, these stories always introduce some stakes, usually through the threat of some great evil which is also pursuing the object. Thus the quest is imbued with a far greater importance, because it becomes not just a search for an item, but a quest to prevent that item from falling into the hands of the forces of darkness. This pattern has existed for millennia, and is largely unchanged from the Greek myth of *Jason and the Golden Fleece* to the twentieth century film *Raiders of the Lost Ark.*[5] Raising the stakes transforms simple tales of the hunt for a missing item into epic missions to save the world.

The concept of raised stakes also appears in stories from other genres. In the typical detective story, where an inspector must solve a crime, the stakes are raised by placing a time limit on the investigation; the inspector must solve the crime before the train reaches the station, or before his retirement, which is only a week away. In the standard farce, where our heroes find themselves in an absurd ruse, the stakes are raised by the threat of something (usually violence or financial ruin) that will befall them if they abandon that ruse. Stories about tortured geniuses struggling against the odds to release their work to the world raise the stakes by showing the genius's inner demons threatening to destroy her, and so on.

The raised stakes may manifest themselves in any number of different forms, but their purpose is always the same; to heighten the emotional tension felt by the audience about the narrative proposition. The stakes are what will happen if the proposition is not fulfilled, if the someone does not get the something they want so badly. They are the answer the the question: "who cares?" The stakes give the audience a reason to invest emotionally in the outcome of the story, and they can be useful in giving an otherwise mundane goal a much greater importance.

Lastly, it is worth noting that while the stakes are a very common element in many stories of many types, high stakes are not universal, and are not *always* necessary. The best example of this is the love story, which is almost always about two people who want desperately to be together, but something keeps them apart. The great love stories rarely do anything to raise the stakes, even though at first glance the struggle of the protagonists may seem prosaic. After all, what happens if our lovers never see one another again? They are not trying to save the world, or to right any great wrong,

---

5 The classic quest, also known as the monomyth, is the subject of one the best books about story, *The Hero With A Thousand Faces*, by Joseph Campbell.

they simply want to be together. The reason love stories work without any great stakes, is that *the thing the characters want is something the audience wants for themselves.* Everyone wants to love, and to be loved, and love that is lost is a tragedy in its own right. Every audience member knows what love is, and why one would want it, and how it feels when it is denied to them, so they are already emotionally invested in the narrative proposition.

That said, such stories represent a minority of the possible kinds of stories that people tell. In any circumstance in which the characters want something (destroy the magic ring, unravel the secret code, construct the perfect bridge) that the audience doesn't necessarily care about in advance, one finds that there is also the threat of some dire consequence if our heroes fail. The device is used more often than not, because it is so effective in forcing the audience to invest emotionally in the story's outcome.

Make the Trouble Hard

One of the problems that beginning writers often face when crafting stories of their own is creating difficult challenges for the protagonist to face in pursuit of the goal. This may because they have taken such care in crafting their main characters and the world they inhabit that they are reticent to make life difficult for those characters. Yet doing so is absolutely necessary; our someone must have trouble getting what she wants, and the harder that trouble is, the better.

The reason this is essential is that the hero's collision with the obstacles that stand in the way of her goal is what causes the audience to empathize with our hero's plight. Recall that the reason the theorem of storytelling works is that human beings have an instant emotional connection to unfulfilled desire. The more trouble our protagonist has in pursuit of that desire, the more deeply the audience will internalize her struggles as their own.

A common technique used in structuring the obstacles for the hero to face is called the "reversal," the point in the story where our hero appears to be getting closer to the goal, before suffering a major setback that leaves her even farther from her objective than when she began. A long story may contain several such reversals or setbacks. The purpose of these setbacks is heighten the emotional tension felt by the audience over the narrative proposition, for the closer one gets to to object of one's desire, the more bitterly one feels its absence.[6]

Perhaps the best example of this process in action can be seen in the course of any major sporting event. Spectators are universally more enthralled by the "close contest," by the struggle whose outcome seems uncertain until the last possible moment. Indeed, we often measure the quality of such contests by the number or magnitude of reversals they contain; the so-called "great game" is the one where the eventual winner "comes from behind," overtaking the opposition only at the end of play, or where the lead frequently changes.[7]

These vicissitudes of fortune are not embellishments or ornaments to story, they are an essential component to story's success. Someone must want something badly, and have trouble getting it. The worse that trouble is, the better.

Victory is Irrelevant

The last principle of effective storytelling is counterintuitive, but simple to implement, provided that one has faithfully applied the earlier principles. Provided that the storyteller has introduced a protagonist,

6 The interested reader should consult the story of Tantalus, a hero of ancient Greek myth, whose eternal punishment by the Gods is the origin of the English word "tantalize."

7 We will return to explore the analogy of sport in more detail in Chapter Thirteen.

established an objective, raised the stakes for achieving that objective, and inserted significant obstacles into the protagonist's path, *it does not matter whether the hero achieves the goal or not.*

To be certain, whether or not the hero accomplishes the objective does determine what type of story our story will be, but a story in which the hero fails in his quest is just as effective a story as one in which he succeeds. Many of the best-known stories are tragedies, after all, where our heroes end up heartbroken, or destitute, or dead. The audience for such a story is likely to close the book or leave the theater with a real sense of grief, but almost never walks away from the story regretting the fact that they heard it told. Here again a useful parallel can be drawn to sport: the spectators who file out of a stadium after their team suffers an embarrassing defeat will certainly mourn the loss, but they seldom wish they had not seen it.

What's more, it is not only perfectly acceptable (and at times desirable) to construct a story in which the hero fails to achieve the goal, it is also perfectly acceptable at the outset to tell the audience how the story will end. This again seems counterintuitive, since we made clear that the purpose of the narrative proposition is to force the audience to invest emotionally in the outcome of the story, but the truth is that audiences are invested in the *process* of story, not in the story's end *product.* We are less interested in *if* the hero gets the girl, or defeats the villain, or locates the treasure, than in *how* that hero goes about pursuing that goal. This is not to say that there is no place for mystery in storytelling; certain kinds of stories like "whodunnits" require the careful concealment of certain information from the audience until the right moment, but that is a question of *how* the inspector solves the crime, not *if* the crime gets solved. The storyteller should not tell us the identity of the murderer, but it's fine if we know that the inspector will expose her eventually.

The resolution of the dramatic tension, the final victory or defeat of the protagonist in pursuit of the objective is something that many stories convey to their audiences before the story even begins. Stories of mythical heroes (e.g. Odysseus, Henry V, James Bond) almost always end in victory, while the subjects of tragedy (e.g. Oedipus, Hamlet, Tony Soprano) almost always meet final defeat. The audience usually knows that before the story begins, but that knowledge does not erode their interest in hearing it. The destiny of the protagonist is an implicit component of the contract made between the storyteller and the audience, and a defining trait of the kind of story we are going to tell.

The tenets we have outlined in this chapter have been drawn almost exclusively from fictional stories, but they apply equally well to stories of fact. In the next chapter, we shall explore the particulars of how they can be used to tell stories about science.

5. Science As Story

ELLIE
They should've sent a poet.

-CONTACT (1995)

Science is, at its core, a search for truth. It is a quest to reveal the underpinnings of the operation of the cosmos. The foundational principle of science is that the truths it reveals should be verifiable independently from the individual observing them. Hence science demands an unswerving adherence to the facts, without exaggeration or embellishment. As we discussed in Chapter Two, this leads many who practice science to view with suspicion any process or product that attempts to convey truth by appealing to emotion, and storytelling is no exception.

This situation is worsened by the fact that in the English language, the word "story" is colloquially used to mean "fabrication" or "lie." There is an oft-repeated American adage that one should "never let the facts get in the way of a good story."[1] Story, then, is a loaded term to many who practice science, for story is inextricably linked to the corrupting influence of emotion, and story has the whiff of the imaginary or the magical. Storytelling seems like a thing the conscientious scientist should avoid, because storytelling

1 Attributed in various places to both Mark Twain and Ernest Hemingway, the phrase is of unknown origin.

feels unscientific.

In this chapter, our goal is to overcome that hesitancy. First, we examine the various kinds of stories that one can (and should) tell about science, by looking at how some of the most common scientific challenges can be framed as narratives, and providing tips on how those narratives can be constructed. Next, referencing the order of operations introduced in Chapter Three, we will discuss how a narrative can be crafted to achieve a specific impact on a specific audience. In the final section of this chapter, we take a wider view of the scientific enterprise as a whole, in order to illustrate some interesting parallels between the roles that science and story play in human culture.

Types of Narratives

The first principle to embrace when approaching the problem of how to tell stories about science is that effective storytelling does not require any alteration, obfuscation, exaggeration or manipulation of the facts. One can construct a crisp, concise, compelling, emotionally engaging, and highly memorable narrative while staying completely faithful to the factual accuracy (and thus the scientific integrity) of the subject under discussion. Many of the great stories through the ages are true, and there is no shortage of narrative storytelling today (e.g. documentary films) devoted to conveying truthful information.

Creating a narrative then, is not a matter of altering the facts, but rather a process of carefully selecting which facts are relevant, and the order in which they will be presented. If we are careful to select the information that most neatly structures a narrative proposition, and we introduce that proposition at the outset, all the rest of the information we present will be viewed by the audience through that narrative reference frame. This creates an emotional investment on the part of the audience, and gives them a reason to lean forward, to engage more fully in the content you are about to present. Here, we will look briefly at some of the most common types of stories one sees in science, delineated by the type of protagonist found in each story.

Scientist As Hero

The most common form of story about science is the story in which the scientist is the story's protagonist. As we've discussed, scientists bristle at this notion, repelled by the idea of portraying themselves as heroic, and uncomfortable inserting themselves into an account of research that is designed to be objective, and free from the influence of any personal agenda. Yet recall that the story's protagonist is not necessarily the "hero" in a classical sense, no mystical powers (or even any profound achievements) are necessary. The protagonist is simply the someone who wants something badly, and who has the trouble getting it.

This form is so common simply because it lends itself to the nature of scientific investigations. We rarely begin observations or experiments simply to occupy our time, we begin observations or experiments because there is something that we want to know, or something we want to show. Implicit in that investigation is a narrative proposition. You, the scientist, wanted something badly, and had trouble getting it. The meat of the story comes from your account of the trouble; the series of false starts, dead-ends, and dashed hopes that are always part of the process of research.

For those who may still be discomfited by the notion of painting themselves as protagonist, it is often useful simply to switch from the first-person singular to the first-person plural: change the "I" statements to "we," and the "me" statements to "us." In practice, this is the way many of us in the sciences speak about our work casually, and thus may feel far more natural. It is perfectly acceptable to tell a story where the protagonist is a group of people (e.g. your lab, your project team, etc.), rather than an individual.

Regardless of whether your protagonist is an individual or a group, it is absolutely essential that you identify who your protagonist is, and what goal he/she/they sought, in the beginning of the story. Recall from the previous chapter that the reason narrative storytelling works is that it engages the audience's emotions, by creating an empathetic response to the protagonist's unfulfilled desire. That mechanism can only operate if the audience is able to identify that desire, and that identification must occur before any account of the trouble the protagonist encountered.

Depending on your audience, it is usually also desirable to introduce some stakes, as discussed in Chapter Four. Generally speaking, the less familiar your audience is with the subject, the more important it is to define the stakes. When speaking to a strictly professional audience of your peers, it is likely that the value and importance of your line of inquiry will be understood, but when addressing a more general public, you should spell out in detail not only what you were looking for, but *why* you were looking for it. Recall from Chapter Three that you should select your content based on the impact you hope to have on your audience. For any audience that is going to hear about something new to them, it is critical to explain the importance of your work, and this is most easily accomplished by outlining what is at stake.

Note that none of this requires any alteration of the facts, merely a careful consideration of which facts to present, and in what order. Telling a good story does not require you to lie. You can tell the true story of yourself and your research, but in order to be effective in must be structured as a story, and that means telling your audience up front what it was you wanted, and why you wanted it so badly.

Non-Human Heroes

There are many circumstances in which it may not be desirable or appropriate for a scientist to tell a story in which she is the main character. In applications for funding, for example, painting oneself as the protagonist risks the appearance of self-aggrandizement, and it may be preferable to find another narrative framework. One option in such cases is to construct a narrative proposition in which the subject of the science becomes the protagonist. This is possible even if the subject is non-human, or completely inanimate, through a technique called anthropomorphization.

Anthropomorphization is the assignment of human traits or characteristics to non-human things. The classical example of this process is found in fiction (often in cartoons), where a creature like a mouse or a dog walks and talks. The concept of anthropomorphization is far broader, however, and is used to ascribe any recognizably human impulse to an object or entity that is non-human. In many cases, it is possible to attribute some desire to the subject of your particular research, and in doing so, create a narrative proposition.

If this notion seems outlandish, it is worth considering that because of the power of human beings to empathize, people anthropomorphize all manner of things, and science is no exception. We casually refer to the seed that "wants to germinate," or the electron "trying to return" to a lower energy level. We talk about viruses being "smart" and refer to computer programs as "clever." In each of these cases, we are assigning human emotions or traits to non-human (in some cases inanimate) systems. We do so without thinking about it, and because our brains are pre-wired to think in those terms, we often find that doing so is an aid to developing a conceptual framework for understanding new phenomena. Because this process comes so naturally, it can be easily applied to the construction of a narrative proposition.

Whether it is a system that wants to return to equilibrium, a species that wants to outpace its competitors, a technology that wants to be adopted, a device that wants to improve a process, or an effect that wants to be fully explained, anything can be framed as the protagonist in a narrative, provided that the thing can be said to want something badly. As with all other narrative constructions, the only requirement is that the audience understand who the main character is, and what that character wants, in the beginning of the story. Our brains are so attuned to narrative frameworks that we seek story in everything we see, and we will gladly accept a story in which the protagonist is a rock, or a star, or an atom, provided that the story is presented in such a way that makes the narrative proposition clear.

Society As Hero

The third type of story that one hears about science is the one in which society at large is the protagonist, and the narrative proposition is our quest to defeat some common enemy. This is the framework that one most often sees when the subject under discussion is extremely broad in scope, or intersects closely with questions of public policy. Whether the topic is global climate change, or nuclear proliferation, or the spread of an emerging epidemic, or the depletion of a natural resource, the statement of the issue implies a narrative, in which we, the members of society, want badly to prevent the terrible consequences that we all agree will result if the issue is not addressed.

The great advantage of this device is that it offers the chance to involve the audience more fully in the narrative than is possible in any of the other forms. For if society as a whole is the main character of the narrative, then the audience listening to the tale, as members of that society, are also characters in the story. This allows the storyteller to offer the audience the opportunity not just to observe the story, but to participate in it, turning them from passive listeners to active voices. Recall from Chapter Two that it is the Dionysian impulse, our emotional brain, that most strongly compels us to act. By creating a narrative structure in which the audience can play a role in the story's outcome, one can fully engage that impulse, which is why this narrative structure is to frequently used by those campaigning for public involvement. By establishing a common enemy against which we the audience must rally, this narrative can fully immerse the listener in the course of events, which causes them to make the fullest possible emotional investment in the story's outcome.

Indeed, the notion of society as hero is the narrative framework that undergirds the scientific enterprise itself. Science is fundamentally a struggle to explain things, to describe, understand (and in many cases, manipulate) the phenomena we observe in the universe. It is a struggle against incomprehension, against ignorance, and the establishment of that enemy creates a narrative proposition in which science itself becomes the hero, who wants badly to know how the world works.

This is the narrative that one sometimes sees used to justify public support for scientific research, particularly basic research. Researchers investigating the kind of fundamental questions that have no obvious or immediate application often struggle to articulate why their work deserves public support. The employment of this narrative is one way to answer that question. Science wants badly to be complete, to fill in the gaps in our knowledge of the universe, because we as a society value that knowledge, and have learned that our ignorance is often injurious. By framing the knowledge acquired through science as a thing that we as a society want badly, issues like funding sources and meddlesome regulations become part of the trouble we have getting it, a framework that encourages the listener to invest emotionally (and, it is hoped, financially) in scientific progress.

It is important to note here, that despite the power and utility of framing all of society as a protagonist, the framework is not without its weaknesses. Chief among these is that because the narrative presumes that the listener wants the same thing the storyteller wants, the device risks alienating those potential audience members who do not share the desire, or who do not subscribe to the same narrative frame of reference. A fuller exploration of this phenomena, which we will refer to as the divergence of frames, appears in Chapter Thirteen.

Choosing Narratives

Which of the three preceding narrative forms you will choose for a given piece of communication should depend entirely on who your audience for that communication will be, and what impact on that audience you hope to achieve. Recall the impact statement that was discussed in Chapter Three. Our aim was to state a specific, achievable, and hopefully measurable impact that we wished to have. This statement is important not only because it helps guide us toward the most appropriate audience for our content, but also because it helps us choose which narrative we will employ to frame that content.

In general, the section of a story that the audience will remember best is what is known as the *denouement*, the final resolution of the story, the answer to the question posed by the narrative proposition. It is therefore best to structure the narrative in such a way that the end of the story somehow establishes or reinforces the impact statement. To illustrate this process, we will present two hypothetical examples of two very different types of scientific communication, and how deliberately chosen impact statements can guide the construction of a narrative.

First, consider a researcher preparing to give a talk at a technical conference. She will be appearing at one of several dozen concurrent symposia, which will feature a series of talks featuring work very similar to her own. The audience in her symposium will be almost exclusively fellow researchers from her field, many of whom she knows personally or has worked with in the past. She is presenting the results of a recent series of experiments which demonstrate the novel application of a particular substance for the improvement of an existing process. This could be a gel for potential use in a gene assay, a semiconductor potentially useful as a photovoltaic, a drug that shows promise as an analgesic, etc.

In this case, her impact statement is simple, she wants her fellow scientists to leave her talk believing (as she does) that her application of this substance is a promising pathway to improve this process. Thus the end of her story must reinforce that point by driving home the notion that this substance has shown promise and (presumably) should be pursued.

Working backwards from this conclusion, we can see a logical narrative frame emerging. She (or her lab group, or team) is the hero, and the thing they want badly is to improve the existing process. Her new application of the substance is a tool that she and her collaborators have used in pursuit of that goal. The process of developing the application and verifying its promise required several laborious steps, which were the trouble that they encountered. The story should begin with an introduction of the goal (improvement of the process), and proceed with a description of the trouble (implementing use of the substance). The story concludes with the goal unobtained. They wanted something badly and did not get it. But they did get closer than they were before, as a result of the novel application of the substance. This is where her telling should end, because this is what she wants the audience to remember.

Now, consider the same scientist, this time invited to give a public lecture at her local science center.[2] Her talk will take place on a weekend afternoon, and she is told that the audience at that time is most likely to be families with young children, though the talk is open to all visitors, and may draw a significant number of undergraduate students from the local

2 More formally known as institutions of informal science education, such centers will be discussed in more detail in Chapter 14.

university.

In this case, her desired impact is less immediately apparent. She cannot select the same impact she desired in the previous case, because the audience is unlikely to be familiar with either the substance in question or the process she is trying to improve. Recall that the impact statement should be phrased as the thing she wants the audience to know, think, or feel after her talk that they did not know, think, or feel beforehand. She has several options. Her chosen impact may be to make the audience believe that her field is a fascinating one where many unsolved questions remain. Or it may be to make the audience feel that her field is a benefit to society, and is contributing to an improved quality of life. She may want to impress upon the audience the importance of her institution, and how it has supported groundbreaking work like hers. Or she may simply be hoping to leave her audience feeling positive about science as a whole, and understanding that science would benefit from the participation of more women and girls.

What matters most is that she select an impact, and that it be specific and achievable. (Few if any impacts in this context are likely to be measurable). For the purposes of illustration, let us assume that her chosen impact is to leave the audience knowing more about her field as a whole, and feeling that it promises to improve overall quality of life in the coming decades. For the sake of clarity, let us assume that her process is the photovoltaic generation of electricity, and her substance is a newly-fabricated semiconductor.

Given the audience for this talk, she can assume no prior knowledge of the science involved. She must assume that basically all of the scientific concepts presented will be new, and as such she must choose carefully which to include, and which to leave aside. Given these constraints, one of her best options would be to structure a narrative in which the science itself is the main character. Her introduction might go something like this:

> The modern world depends on energy. It grows our food, operates our cities, drives our systems of transportation, and powers our bodies. Humans need energy that is cheap, clean, and plentiful. That energy comes to the earth in massive quantities from the sun, but for many years we had no effective way to capture it. Until the emergence of a new and powerful technology: the photovoltaic.

Notice how, in just five sentences and 66 words, she has introduced the subject of her talk, and at the same time, framed a clear narrative proposition. Almost instinctively, the audience will anthropomorphize the technology, identifying with it as a character in the story. Someone (the photovoltaic) wants something badly (to provide the world with cheap and clean energy from the sun). By introducing an objective, and establishing a protagonist who will strive toward it, she has established a narrative frame.

From this simple introduction, the remainder of the narrative neatly emerges. The photovoltaic wants to create abundant, efficient energy, but there are complications (the bandgap, the re-combination of exciton pairs, etc.) that prevent it from doing so. Every other piece of information that she presents is given a greater weight, an emotional importance, because it now appears in the context of the narrative. Her story will progress through an explanation of the trouble the photovoltaic faces (mentioning the work that goes on in her lab as one of the places where a small victory is being won) and conclude, as in the previous example, with the goal still elusive, but closer to being obtained than it was before.

Again, notice that nowhere in our example did our scientist embellish or exaggerate the facts. The art of storytelling does not require the fabrication of information, it requires only a careful consideration of the audience's response when ordering the presentation of extant information. Narrative framing can be accomplished with minimal effort, provided that one begins with a clear impact one hopes to have on a specific audience.

Nested Narratives

To conclude our discussion of the employment of narrative structure, we consider the concept of narratives within narratives, or nested narratives. In the final example above, there were in fact several narrative propositions operating simultaneously, each within the other. The largest narrative was about the world (someone) wanting energy (something). Within that frame, there was the photovoltaic (someone) wanting efficiency (something). Within that frame, we had the researcher (someone) wanting to improve her particular semiconductor (something). The frames stack neatly together, each existing as an independent within its larger, or parent narrative. We see this represented schematically in Figure 5-1.

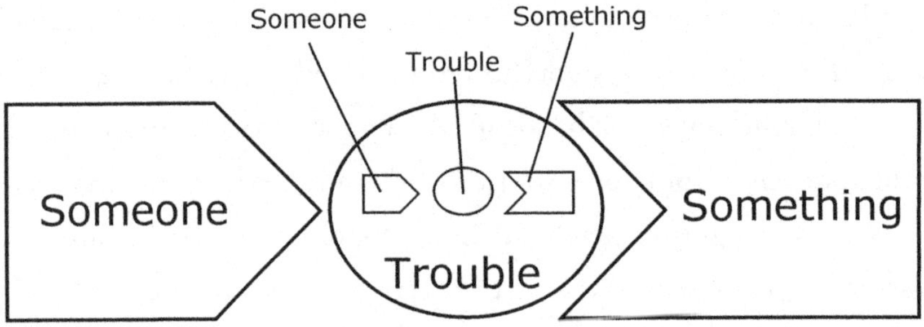

Figure 5-1: Nested narratives. Within the trouble we find another someone and another something.

This nesting of narratives one within the other is a common device, and is nearly ubiquitous in fiction. Within the larger narrative of the film or novel or play, there exists several sub narratives, (the acts) which are in turn broken up into constituent units (the scenes). Because narrative is so ingrained, and because our brains are seeking narrative everywhere, the existence of this structure may be invisible to the average audience member, but it is almost always present.

While developing nested narratives is not strictly necessary, it can be a useful concept, particularly when communicating with audiences who may not be familiar with the subject at hand. By starting from a narrative frame that the audience is familiar with (e.g., the world's need for cheap, clean energy), you can place the narrative proposition of your work within that existing frame, thus connecting your work to the emotional investment the audience has already made in the larger frame. The reality is that the majority of scientific investigations fit neatly into a relatively small number of existing narrative frames, and one can profitably exploit those frames in a discussion of one's work.

Through the course of the last five chapters, we have discussed the importance of communication in the conduct of modern science, and have presented a theoretical treatment of the framework through which good communication can be most effectively accomplished. These principles are equally applicable regardless of the medium through which the message will ultimately be conveyed. Indeed, a well structured narrative will apply equally well across a broad range of different communication platforms, allowing the skilled communicator to offer a consistent message across a diverse set of mediums.

In the next section, we will leave aside these theoretical considerations, and explore the practical elements of communicating through various pathways. Yet the guiding principle remains; we must have a clear sense of our narrative in hand, as we apply our theoretical framework to practical matters.

Part One of this book was about finding a story to tell. Part Two is all about how to tell those stories.

Part Two

Practice

6. Speech

HOLLY
After all your posturing, all your little speeches, you're nothing but a common thief.

-DIE HARD (1988)

Public speaking (which we can define as speaking when more than one person is listening), is one of the oldest and most important modes of human communication. The tradition of speaking to groups of people is integral to human society; it forms the basis of much of our system of education, plays a central role in our system of government, and for most of human history was the primary method through which knowledge was disseminated and preserved. Science has its own rich legacy of public speaking, which stretches from the Lyceum of Aristotle through the societies of the enlightenment to modern scientific conferences today. The spoken word is probably the most important form of human communication, because (as we shall see) it allows for multiple channels of communication to coexist.

Despite its historical pedigree, speaking before an audience is not something everyone aspires to do, or particularly enjoys. Indeed, for a significant number of people, the prospect of speaking publicly inspires an acute anxiety, if not an outright sense of dread. Worse, very few of us trained in the sciences have had the time or the opportunity to study or practice the art of public speaking, and thus find ourselves called upon to stand and speak before an audience with little or no training in how to go about it.

In this chapter, we discuss some of the fundamental principles of public speaking. Like all of the chapters in this section, the lessons presented here are not meant to improve the reader's communication skills in isolation; instead the principles herein are presented in the hope that the reader will employ them in his own practice at communicating, and use them to refine his own technique.

We begin with a discussion of both the physics and the physiology of the fundamental tool of public speaking: the human voice.

Physics

The sounds that we perceive as speech (indeed all the sounds that we perceive, regardless of their source) are vibrations in the fluid that surrounds us.[1] These vibrations travel through that fluid as transverse pressure waves, propagating outward from their source at a speed determined by the fluid's density and pressure. For our purposes, these waves have two variables of importance, frequency and pressure.

For simplicity, it often helps to represent these waves as sinusoidal plane waves, as shown in Figure 1, where the y-value represents the pressure in the direction of travel of the wave.

1 The explanations in this chapter are, of course, simplifications. A fuller explanation (including the relevant equations) can be found in almost any elementary physics text.

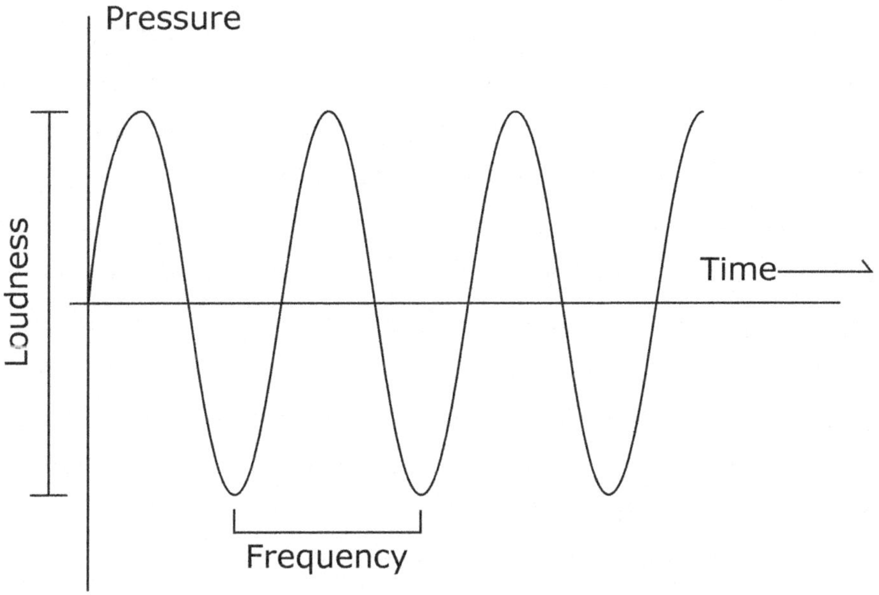

Figure 6-1: A simple sound wave. The period of the wave is what we perceive as pitch, and the amplitude is perceived as loudness. Neither depends on the volume of air being moved.

The frequency defines what humans perceive as "pitch", the highness or lowness of the sounds, while the amplitude of the oscillations in pressure are perceived as loudness.

Human beings can perceive sounds over a wide range of frequencies, from roughly 20 Hz to as high as 20kHz, but almost all human speech (and not coincidentally, most musical melodies), fall in the band between 100 and 1000 Hz. Human perception of frequency is geometric, with a precise doubling of frequency corresponding to one "octave" on the musical scale. (e.g. both 220 Hz and 440 Hz produce the pitch referred to as "A".) The frequency of sounds is preserved as they travel through the medium.

The measurement of loudness is more complicated, as sound pressure varies over a wider range, and our perception of it goes logarithmically. Loudness is most frequently measured in decibels (dB), representing a gain over a certain reference pressure that is generally considered the threshold of human hearing. The amplitude of sound waves decays as they travel away from the source, in a manner that can be calculated from first principles, but because the human hearing system responds non-linearly, there is not a simple way to calculate perceived loudness based on distance alone.

What's more, sound waves impact objects and materials in the environment, and can be either reflected or absorbed, depending on the material properties of the surface. As a result, the degree to which loudness will fall off with distance varies immensely depending on the space in which one is speaking, which is why acoustic considerations have long been a key component in the design of public spaces.[2] For our purposes, there are two key principles of the physics of sound.

First, it is important to remember that the particles of the medium do not travel with the wave. This is intuitive for the propagation of sound in a solid, but is less obvious when we are creating sounds in air with an air-powered instrument, like a trumpet or the human voice. In those systems, air is used as the motive force to vibrate some physical object, but the propagation of the sound does not rely on the air moving away from the instrument. A violin plays just as powerfully as a clarinet, though it is driven without any air at all.

Second, at the loudness of a sound (and thus the ability to distinguish individual words) is a function of the *pressure* of the sound wave. This can be confusing, because loudness is so commonly referred to as "volume". The two must be kept separate, because *volume* (in the real physical sense, i.e. the quantity of air), is a useful parameter in describing respiration, which is a critical component of speaking. As we shall see in the next section, efficient speech requires releasing air in small volumes and at high pressures,

2 Examples from antiquity, including the Colosseum in Rome and the Theater of Dionysus in Athens, allowed an unamplified human to be plainly heard by tens of thousands of listeners.

to produce a loud, clear speaking voice.

Physiology

The human voice can be thought of a pneumatic machine, with three main components. First is the respiratory system, in which the lungs act much like a bellows to drive a column of air. Second is the larynx, in which the vocal cords (also called the vocal folds) vibrate in that column of air to produce sound at a frequency determined by the tension exerted upon them. Last is the mouth, which uses combinations of lips, tongue, and teeth to form the consonant sounds that make up the phonemes that form the constituent parts of words.

The most important part of that process is ventilation, the act of breathing that powers the whole pneumatic system. Ventilation is powered by two cooperative muscle systems: the intercostal muscles and the thoracic diaphragm. The intercostals are muscles that lie in between the individuals ribs, while the thoracic diaphragm is a large band of internal muscles that separate the thoracic cavity from the abdomen. These two groups work together to expand the chest cavity and draw air into the lungs (inhalation) and compress the cavity to expel air outward (exhalation), but they are not equal partners, and one can control which of the systems is the dominant player.

Breathing is both more efficient and more powerful when it is driven primarily by the diaphragm, and not by the intercostals. To understand why this is so, it is helpful to think of the chest as a cylinder made of some compressible material (a large plastic bottle is a reasonable approximation). The intercostal muscles cause ventilation by pushing and pulling against the chest wall, not unlike squeezing pulling apart the the sides of the cylinder. The diaphragm, by contrast, works by compressing the contents of the abdomen, which lowers the floor of the chest cavity, akin to pulling downward on an airtight piston within the cylinder. For this reason, diaphragm breathing uses less energy, allows the use of a greater percentage of the lungs' total capacity, and creates a greater positive pressure within the lungs.

Diaphragm breathing is not difficult, but it is not most people's natural tendency. As an introduction to the practice, it is helpful to develop an awareness of the muscle itself. Here we present two simple exercises that can help you feel the diaphragm in action.

> **1.** Sitting comfortably or lying down on your back, place one hand on your stomach, and with your lips closed, inhale through your nose. You will feel your stomach begin to rise. Continue inhaling until your stomach stops rising. That's the diaphragm. Exhale and repeat.
>
> **2.** Take a deep breath in, pause for a moment, then exhale. Continue exhaling. Exhale some more. And more. Push as hard as you can. Drive as much air as possible can out of your lungs until you can't possibly move any more. Concentrate. Keep exhaling. You should feel a tightening of your abdominal muscles. Now inhale, and you will feel the diaphragm contracting. Try this exercise again, this time blowing into a straw submerged in a glass of water. Blow bubbles until you cannot blow any more. You will feel the diaphragm contracting.

Hopefully, through these exercises, you have begun to feel the diaphragm in action, but it is important to remember that the muscle does not just drive exhalation, but inhalation as well. In normal breathing at rest (sitting at your desk say, or reading a book), the volume of air that travels in and out with each breath is roughly 500 mL, at a rate of somewhere near 20 breaths per minute. This represents a tiny fraction (~10%) of the total volume of our lungs. Through conscious use of the diaphragm, we can increase that volume several fold, and allow ourselves to efficiently move a vastly increased volume of air without increasing our rate of respiration.

This deep, efficient breathing is advantageous for several reasons. First, using the diaphragm and breathing deeply allows us to create a higher positive pressure in our lungs, which means that when we are releasing air, it comes out in bursts that comprise a small volume of air at a high pressure, which is what determines how loud the produced sound will be. Second, fuller breathing means that we will vastly increase the number of words per breath that we are able to deliver, which gives us greater latitude regarding where in each sentence we wish to pause for breath. Lastly, this deep, slow breathing is a more efficient form of respiration, delivering an abundant supply of oxygenated blood to the muscles and the brain.

Deep, purposeful breathing from the diaphragm is also an important component in perceptual awareness, and in controlling the responses of the nervous system, particularly in circumstances of high stress. Indeed, rapid, shallow, chest-driven breathing, or panting, is a symptom of panic, exhaustion and shock, and is almost always a sign of something gone terribly wrong. Athletes, adventurers, musicians, meditators and martial artists all concentrate closely on slow, purposeful, diaphragm-driven breathing, because centuries of experience have proven how effective it is.

Perhaps the best preparation for any instance of speaking publicly is to concentrate on slow, deep, efficient breathing. This will lower your heart rate, oxygenate the blood flowing to your brain, and straighten your posture. When you are ready, face your audience, fill your lungs, and begin. Your voice will come out clearer, louder, higher in pitch, and you will find that whatever nervous energy consumed you before you began has already begun to dissipate.

Enunciation

Speech requires more than just the use of the vocal cords to produce tones, it also depends on the use of the lips, mouth, teeth and tongue to create the consonant sounds that define words. These consonant sounds are especially important when speaking publicly, because as sound waves travel, their decreasing amplitude makes them less distinct from one another. This is obvious from experience: we have all heard people nearby whispering or speaking quietly, and though we know for certain they are speaking, we cannot make out the individual words.

When speaking publicly then, it is critically important to enunciate, to form clear and distinct consonant sounds that allow the listener to discern the constituent words. This is particularly important when speaking in a language that is not native to either the speaker or the audience (which is very often the case in a modern scientific context). Luckily, proper enunciation is not difficult.

The most important factor in clear enunciation is simply how far one opens one's mouth. The greater the path of travel of the jaw when speaking, the more distinct the consonant sounds are from the vowel sounds. Speakers who appear to mumble are often simply speakers who are not moving their jaws to open their mouths, which tends to blur the sounds together, and make it difficult to tell where one word ends and the next begins. What's more, opening your mouth widely while you speak tends to slow the rate of speech, which is another key factor in making yourself understood.

Proper enunciation can easily be observed by watching yourself speak in front of a mirror. First, face yourself, and try speaking a simple sentence out loud. (It can be any sentence you like). Watch how your mouth moves. Now, speak the sentence again, but try to move your jaw as little as possible. You should hear how the words are much more difficult to discern, and the consonant sounds less distinct. Now, speak the sentence a third time, but enunciate as clearly as you can. Exaggerate the sounds that compose each word. You will see that you are naturally opening your mouth more.

When practicing this technique, it is sometimes helpful to separate the individual syllables in each word. A simple speaking drill is the recitation of three words: "Exaggerate, Articulate, Enunciate." But try saying them like this:

"Ex-agg-er-ate. Ar-tic-u-late. E-nun-ci-ate."

You will see that your mouth moves more in the act of articulating the words.

Countless simple drills such as these exist, and a search will yield myriad other examples. Pick two or three that you like, and memorize them. When preparing for your next speaking engagement, warm up your vocal cords with a little humming or soft singing, and then warm up the muscles of your mouth and tongue by repeating the exercises. Speaking is a physical activity like any other, and muscles that are properly warmed up will perform better and fatigue more slowly.

Emphasis

As mentioned in the introduction, the power of the spoken word lies in the fact it allows a speaker to convey multiple messages simultaneously, and utilize more than one channel of communication. One of the ways this is accomplished is the use of emphasis.

When we speak, we put extra stress on certain words, by increasing the volume at which they are spoken, or putting slight pauses after them. These stresses add emphasis to the stressed word, and the choice of emphasis can alter or augment the meaning of the sentence. Consider a simple sentence like the following (read the sentence aloud):

"There is nothing for you to see here."

Chances are you read the sentence with roughly equal emphasis on each word, and the meaning a listener would have received is the same meaning that would come from reading the sentence on the page. Now try reading the same sentence aloud again, but put extra emphasis on the word in bold.

"There is nothing for **you** to see here."

Do you see how the meaning has changed? We added an extra layer of information into the sentence, without adding or subtracting any words. Spoken in the above way, the sentence clearly implies that there is in fact something to see, but it is not appropriate or not allowable to the listener specifically. Imagine yourself as the listener hearing that sentence. In the first instance, you might assume that there is simply nothing to see, but in the second instance, you intuitively understand that while there is something to see, you personally are barred from seeing it. As you can imagine, the way you feel after hearing that sentence will be different in each of these cases. Consider the sentence spoken a third way:

"There is nothing for you to see **here.**"

Now, again, the meaning of the sentence has changed. It is apparent there is something to see, and that you the listener should probably see it, but this is not the proper place. As a listener, you feel invited in, drawn to the mystery, because the speaker has intimated that she has something to show you.

The point to remember here is that we have not changed the construction of the sentence at all. The words are exactly the same, and in exactly the same order. Yet merely by choosing which words to emphasize, we were able to transform the meaning of the sentence, and more importantly, alter the impact on the listener. Attention must be paid to where emphasis is placed, because if we ignore it, we run the risk of creating layers of meaning that we did not intend.

Inflection

Another important channel of communication that is present when words are spoken aloud is what is known as inflection. Inflection is the emotional timbre, the mood or tone that the speaker strikes when the words are spoken.

Inflection is not an easy thing to define, but it is something that humans are acutely able to perceive. Through subtle changes in pitch, in the cadence of words, in pronunciation, etc., a speaker can completely transform the meaning of the message from what is suggested by the words alone. As an example, consider a simple commendation.

Imagine you just watched a colleague or a friend accomplish some difficult feat, be it a precise shot in billiards or golf, or a clever ploy in a chess game, or any well-executed demonstration of skill. You might say:

"Nice Work!"

The listener can tell from the sound of your voice that you are sincere, that are congratulating him on a job well done. But that's not the only way to say those two words.

Now imagine that you had just watched a colleague botch something terribly, dropping and breaking a tray of glassware, or backing into a parked car, or some other embarrassing failure. You might turn to that person and say:

"Nice Work!"

Try saying the phrase out loud, first imagining yourself in the first situation, and then imagining yourself in the second. You can probably hear that the delivered meaning is diametrically opposed in the two cases, despite the fact that precisely the same words are spoken. The difference between the two is one of inflection.

This example (sarcasm) is probably the most extreme case, but inflection can vary in much subtler ways. Because human speech is so deeply ingrained in our psyches (it predates writing by tens of millennia), our cognitive architecture is evolved to detect meaning in speech independent of words. If you have ever had the experience of travelling in a country where you spoke little or none of the native language, you have seen this process in action firsthand. It is possible to discern a great deal about the speaker's intent (and certainly about his emotional state), despite a complete ignorance of the meaning of the words themselves.

Attention must be paid to inflection when speaking professionally, because if it is ignored, the speaker runs the risk of delivering a message she did not intend. In general, the scientist speaking about her work usually wants to appear honest, forthright, trustworthy, and sincere. Your openness to new ideas, the quality of your thinking, and the purity of your motives are all things that your audience will weigh in emotional terms, and thus the mood and tone of your speech is important, independent of the words themselves.

Body Language

A third communication pathway existing in spoken language that is absent from the written word is body language. Body language is probably the most basic and fundamental mode of human communication. Indeed, body language forms the basis of communication between species, as anyone who shares their home with a canine or feline companion will attest. Because body language is so elemental, we react to it without thinking. Our brains are hard-wired to make certain assumptions about other creatures based on their body postures, and that circuitry completely circumvents any rational thought. Thus it is necessary to be mindful of how we carry ourselves when we speak publicly, in order to avoid delivering an unintended emotional message.

To very good first approximation, people judge the trustworthiness and sincerity of people based on how vulnerable they appear to physical attack. Crude though this may seem, it is the result of several million years of evolution, and it is still with us today. A speaker who is exposed, by standing up straight, keeping a wide stance, lifting the chin, pulling the shoulders back behind the plane of the breast, and keeping the hands low, inspires in the audience a sense of confidence and trust. It makes us feel like the speaker has nothing to fear.

Likewise, a speaker who assumes a defensive posture, crouching in a stance, lowering the chin, bringing the shoulders forward, and the hands in front of the chest or face, seems like someone we cannot trust. We intuitively feel that the speaker has something to hide, and is either sinister or a sycophant.

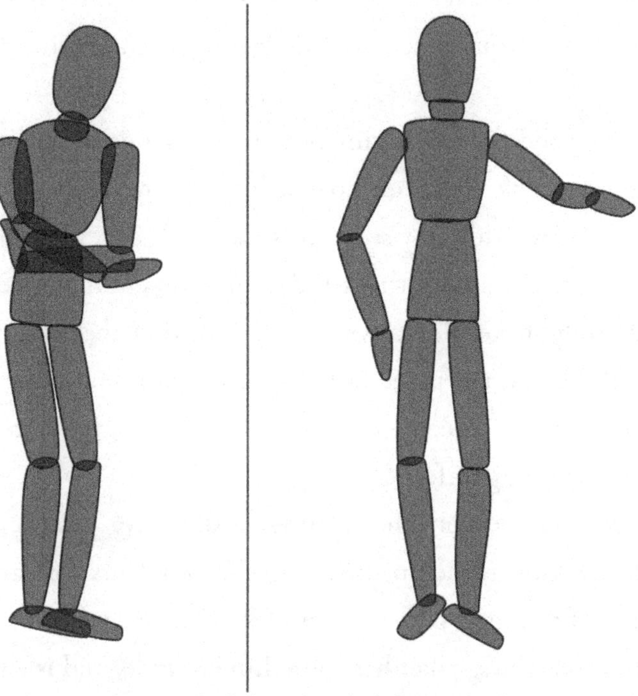

Figure 6-2: On the left, a closed, defensive body posture, which makes a speaker seem untrustworthy. On the right, an open posture, which conveys confidence.

Again, it is important to remember that this is all happening through a very low-level processing center in the brain. If you polled the audience as to how they felt about the two speakers, it is unlikely that most of them would be able to articulate why they found one more trustworthy than the other, but there absolutely would be a measurable difference between the two.

The spoken word is a powerful tool, as it allows a speaker to do what is referred to in signal processing as multiplexing, in which multiple communication signals are sent over a single medium. Through proper breathing, good projection of the voice, enunciation, emphasis, inflection, and body language, a speaker is able to layer several additional signals into speech beyond the meaning of the words themselves.

Before we close this chapter, it is worth noting that every one of the pathways described herein operates effectively only when the speaker is facing the audience. If the speaker is turned away from the people to whom she is speaking, she is harder to hear, harder to understand, her body language is harder to read, her emotional intent is less obvious, and the audience intuitively feels that she cares less about them than she does about whatever else she might be looking at. This fundamental principle of public speaking is so intuitive that texts on the subject from two or three generations ago never even bother to mention it. Yet in the last several decades, a common practice has begun to emerge that makes it imperative to state this edict explicitly:

A speaker must face his or her audience.

The reason we must state this so clearly stems from the use (and common misuse) of one the staples of scientific and technical communication, and the subject of our next chapter: slides.

7. Slides

```
                    SPOON BOY
...try to realize the truth... There is no
spoon. Then you will realize it is not the
spoon that bends, it is only yourself.

                            -THE MATRIX (1999)
```

In modern practice, the scientific or technical talk given before an audience rarely if ever relies on the speaker's words alone. It is now standard practice (almost always expected, and at times required) for the speaker to employ some form of visual aid, the most common of which are the digital images commonly referred to as "slides." Use of these images has proliferated in the past two decades, and the availability of software to create and display them has made them nearly ubiquitous at all levels of science and science education.

In this chapter, we will address some of the most common misconceptions about effective use of slides. We will discuss the history of how and why this form of visual aid developed, and working from first principles, present best practices for the design and use of these visual aids, including choosing the most appropriate software.

Nullius in Verba

In November of 1660, a group of men met in London and founded an organization for the promotion of learning through experiment. The men were at the vanguard of the period of history known as the Age of Enlightenment, and they embraced an idea that was new to society at the time. These men believed that the laws that govern the phenomena we observe in the universe could be discovered and described by human beings, and that the way to discern the validity of these laws was through observation and experiment. That idea, radical in its day, forms the backbone of what we now call science, and their organization, the Royal Society, is now the oldest scientific society on earth.[1]

The Royal Society began a series of weekly meetings, during which its members would present the results of recent experiments, designed to demonstrate some facet of the new scientific knowledge. Soon after its founding, the society also created a publication, *Philosophical Transactions*, which established the model of the scientific journal that is still in use today. That the conventions for sharing scientific knowledge established by the Royal Society have survived for 350 years is a testament to their efficacy. Yet even more important than the means of knowledge transmission the society embraced is the attitude they took toward the way that knowledge was created.

The society chose as its motto the Latin phrase: *Nullius in Verba*, which translates to "Nothing in Words," or more accurately, "On No One's Word." They chose this motto because it neatly encapsulated the fundamental tenet of their beliefs; that they would not accept the truth of any argument based solely on the authority of its author. This flew in the face of established philosophical convention of the time, which held that ultimate authority rested in the texts of ancient scholars like Aristotle, or in the words of the Bible. The men of the Royal Society disagreed. They demanded that one show proof of one's arguments, and they published a journal and hosted a

---

1 An enlightening view of the Royal Society and its impact on the course of science can be found in *Seeing Further*, edited by Bill Bryson.

series of lectures to provide a platform where that proof could be presented.

That notion undergirds all of modern scientific investigation, and remains the fundamental principle upon which all of modern science rests. 300 years after the founding of the Royal Society, in 1966, the American physicist Richard Feynman was asked to define science, and he offered that science is "a belief in the fallibility of the experts," a neat recapitulation of *Nullius in Verba*. Science demands evidence, and the scientific tradition is that we *show* that evidence, rather than just describing it.

Modern science in the 21st century is heir to those principles established in the 17th. In our universities, in our professional societies, and in laboratories throughout industry, government, and academia, the tradition of the presentation of scientific results by the scientist who generated them persists, and the reason for it is unchanged. At our largest gatherings, scientists from all over the world still congregate to hear one another share their most recent findings. What has changed is the substance of what is presented. Gone are the days when a researcher performs an experiment before an audience, and what we see today is a presentation of an experiment's results. And in nearly all cases, those results are presented within the images we call slides.

There Are No Slides

It is worth noting at the outset that despite the language that is nearly universally employed to describe them, modern scientific presentations do not actually use "slides." The term originally referred to pieces of celluloid film held in cardboard or metal holders, which were kept in a carousel or magazine. The pieces of film could slide into the plane of projection between a lamp and a lens, and thus be projected on a screen. The slides were created using photographic techniques, and were static images that were presented a single frame at a time, in a pre-defined linear order, because that was the limit of the technology that governed them.

Beginning in the 1980s, computer software emerged that could create these images using a digital rather than a photographic process, and eventually the slides themselves were replaced by an all-digital workflow, in which the images created in the computer were digitally projected or displayed on a screen. Modern slideware programs like Microsoft's PowerPoint or Apple's Keynote handle both the creation and projection of these materials, which are for historical reasons still referred to as slides, despite their complete departure from their technological ancestry.

This technological change is noteworthy for two reasons. First, because the most common uses of these images continues to emulate their original use, despite having been released from their technological constraints. Most presenters continue to regard these as static images that must be shown one at a time, in a pre-defined linear order. Worse, some in the sciences continue to use these images as the ruler by which the duration of a talk should be measured. Thus you will hear statements such as: "prepare a short talk, no more than ten slides." This slavish devotion to outmoded convention, perhaps more than anything else, is the root cause of many an ineffectual presentation. Before you begin to design or use images that will be used in a presentation, remember that they are just that: digital images. There are no slides.

On Screen Size

One of the first things to consider when designing an image that will be shown to an audience is the apparent size of the image that will be displayed. We say "apparent" size, because the important factor is not the absolute size of the image, but fraction of the viewer's field of vision that the image occupies. A ten meter movie screen looks much smaller from the far end of a football field than a smartphone screen held directly in front of the face. The apparent size is also referred to as the angular size, and it is measured in degrees of arc.

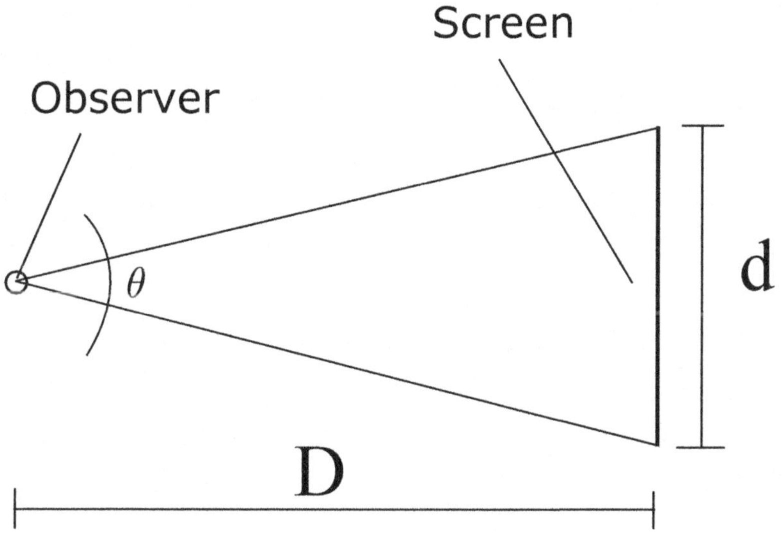

$$\theta = 2\ arctan\left(\frac{d}{2D}\right)$$

Figure 7-1: The formula for angular size. A viewing angle of 30° results when D ≈ 1.87d, e.g. 9.35m away from a screen that measures 5m across.

Long research and experimentation in the motion picture industry has shown that optimal angular size for viewing screens is between 30° and 40° of arc. A presentation which includes both a speaker and screen should therefore roughly divide this in two, affording 15° to 20° of arc each to both the speaker and the screen. In practice, this may be difficult to achieve, and it is seldom the case that an invited speaker will have any control whatsoever of these dimensions. Nonetheless, they are worth noting, if only to highlight how the inadequacy of the projection technologies used in many technical settings. Many technical talks are hampered by undersized screens.

What's more, there can be no useful rules or guidelines about the use of text or figures on slides based solely on the size of the font, as some authors suggest, since the final visibility of the material depends entirely on the size of the projected image, and the relative distance of the observers.

Negative Space

If there is one guiding maxim in the design of effective images (or for that matter, the design of almost anything), it is to include only that which is essential, and omit all that is superfluous. This is not a mandate for minimalism; it does not require the images to feel bare or spartan, it merely insists that every component in the image serve a clear purpose.

This principle can be profitably applied at the very outset, when one chooses the basic color scheme for one's images. In much the same way that a painter starts with a blank canvas, it is useful to consider beginning the design of an image by thinking of a blank screen, and planning to paint upon that screen with light. Considered from this perspective, it immediately becomes apparent that the most natural background for a projected or displayed image is dark, either black or nearly so, the natural color of negative space when a pixel of the screen or projector is switched off. Thus, in order to maximize contrast, the natural foreground color is a light color, like white.

This color scheme, white text and figures on a black field, is counter-intuitive to many, for at first glance it looks inverted from what we have come to expect. The reason for this is that many software programs (e.g. word processors) were designed to emulate the look and feel of working on a manual typewriter, where black ink was laid down on white paper. Yet in digital space, this is in fact an inversion of the natural color space. If you turn your screen off, it's black, and the light areas are places where we are indicating information.[2]

The natural color scheme for a displayed image in a darkened room is light text and figures against a dark background. Projecting or displaying a full white screen, and then shutting off the pixels where we wish to show information is as nonsensical as printing a sheet of paper fully covered with black ink, where only in the letters does the white of the paper show through. (Indeed, if photons were as expensive as ink, it would be just as uncommon). For those who might doubt the wisdom of this advice, it is worth considering the cinema, which has been in the business of projecting images to rooms full of people for well over a century. The end credits of nearly every film are large blocks of projected printed text, and they are always white type on a black screen, rather than the other way around.

Design

As we noted in the introductory chapters, common practice in scientific communication is not necessarily indicative of best practice, and nowhere is this more evident than in the design of slides. Much of the blame for this can be laid at the feet of the designers of the most popular slide software. Since their earliest days, these programs offered the user a limited set of pre-defined templates, sometimes called "master" slides. Because these layouts were so readily available and easy to use, they became the predominant aesthetic for the design of slides, despite the fact that they were not designed with scientific or technical presentations in mind. That said, these programs can produce very effective images, provided that a few simple design practices are followed.

First, it is imperative to remember that these images are visual aids. Their purpose is to show visual representations of information that would

2 It is not a coincidence that some software programs, including several of the most popular suites for software development, now show white text on a dark screen, which reduces eye strain over long periods.

be very difficult to describe verbally. The description of what the listener is seeing should come from you, the speaker. The purpose of the slide is to provide the data or information that supports your argument or enhances your story, not to make the argument or tell the story on its own. A schematic that shows the design of your apparatus is an excellent use of an image, but a list of the advantages of that design is not. The guiding principle here should be (as always) to serve the larger narrative. You are the storyteller, and you should do the telling. The slides should be designed to enhance and enlarge the story being told by showing its aspects (either the someone, the something, or the trouble).

Second, each figure should occupy as much space on the slide as possible. Except in cases where you have two or more images which the audience must be able to compare against one another, this means giving each image its own slide. For example, in cases where you ran the same experiment five times under a set of five slightly different initial conditions, it is perfectly appropriate to show five resulting plots together, but in cases where you want to show two different experiments that demonstrated different effects, give each plot its own slide.

Lastly, explanatory text should be used sparingly and with great care. Printed text is not visual information, it is verbal information, and it should properly be presented by the speaker, and not on the slide. Of course it is perfectly appropriate to use labels that identify the constituent parts of a diagram, and customary to use titles to introduce and conclude your remarks, but the job of explaining what the figures show and why that information is important should be done by the speaker, and not by the slide.

For complex figures that will be the subject of a lengthy discussion, it may be advantageous to include a line or two of text which reinforce the main point. These lines should be as brief as possible, so as not to distract the viewer from the visual information on the slide, and can be thought of as "headlines," akin to those found in newspapers.[3] The key component of these lines is that they always contain *verbs*, the words that tell us what is *happening*. If you find yourself including a line of text that does not contain an active verb, it can likely be eliminated.

It should be noted here that in some cases, in some settings, speakers are encouraged by their colleagues, mentors, or supervisors to use more explanatory text on their slides, at times to the point of nearly transcribing their entire talk. This encouragement almost always stems from the fact that those colleagues have difficulty understanding the speaker, usually due to a barrier in pronunciation or an incomplete mastery of language. This practice is as ineffectual as it is misguided. An incomprehensible talk will not be made better by the addition of incomprehensible slides. If a speaker has difficulties which make it impossible for him or her to present information orally, he or she should not be forced to speak, and should instead submit information through written reports, or some other means. Forcing bad practice in service to custom benefits no one.

Finally, it should be noted that the best slide design will result when slides are designed from the ground up with their final purpose in mind. Using the templates or "masters" included in slide programs means abdicating responsibility for many of the most important design choices. Starting with a blank slide and including that information that seems essential, omitting that which is superfluous, will often yield better results.

As noted during our discussion of body language in the preceding chapter, audiences will make assumptions about the quality of the work based on the visual appearance of the slides. This happens automatically, and often without the audience's awareness, but it can significantly impact the way the talk is perceived. Making slides that look good is not a question of vanity, it is an element of ensuring a successful talk.

---

3 A more complete discussion of headlines appears in Chapter Ten.

Figures

The primary motivation behind the development of visual aids for scientific talks, and still their most important function, is to show data figures. Conventions for the use of figures in research vary widely from field to field, and the most effective employment of figures to convey data (especially statistical data) is a question that continues to evolve. As such a complete discussion of the scientific figure is beyond our scope here, and is a topic that can and does fill several volumes.[4] Here we will highlight a few simple principles that should be noted when preparing a figure for use in a public talk.

First, figures presented during a lecture will only appear on screen for as long as the speaker sees fit to discuss them, and as such should only contain as much information as the audience is able to process in that period of time. A data figure can be an incredibly rich source of information, and when preparing figures for publication, it is appropriate to include the full breadth of relevant data, as the reader will be able to consume that information at her own pace, referencing the figures against the explanatory text.

During a talk, the audience will not have that luxury, and as such figures should be carefully chosen to show only the most important or relevant results, rather than to display the full breadth of gathered data. For example, if you have run ten versions of an experiment that produced ten similar figures, it is perfectly acceptable to briefly show the table, before proceeding to a single enlarged figure, and discuss the trend in that figure that is representative of the entire group of figures.

Similarly, figures which were generated as part of an effort to verify the validity of a result (e.g. the computational agreement with an

4 One of the most comprehensive (and by far the most visually engaging) is *Visual Strategies* by Felice Frankel & Angela DePace.

experimentally derived value), while appropriate in a publication, may be more information than an audience has time to digest, and can be mentioned rather than shown.

The guiding principle should be to show the data that supports the underlying story. Someone wanted something badly and had trouble getting it. If the figure shows the trouble, it should be part of the story. If it does not, it should likely be left aside.

Second, when designing a figure for use in a presentation, it should match the color scheme used in the image or "slide." If (as is suggested) you are using images with dark backgrounds and light foregrounds, this means that the figures should employ this same color scheme. Obviously, figures that are prepared for printing or publication will use a black-on-white color scheme, and should be transformed for use in a presentation. This can be relatively easily accomplished in many software packages by inverting the color map, (often using a function in the program called "invert"), though this will also have the often undesirable effect of inverting colors to their complementary color (e.g. red becomes green, and vice versa). It may therefore be necessary to manually transform the background and foreground colors, a relatively straightforward process in any full-featured illustration program.

While this process can be time and labor intensive, it generates figures of a superior quality. It may not be possible to do this transformation for every talk, but when creating figures that will be used in many talks or by many people, or when creating figures for a talk where the stakes are high (a job interview, a product presentation, a grant review, etc.) it is well worth the effort.

Lastly, when creating a figure for use during a presentation, it is essential to maximize the area of the screen which displays the most relevant information, namely the data. Because the speaker will rarely have control of the real or apparent size of the images being displayed, care should be taken to show the figure at the largest size possible, in order to ensure it is legible by the audience.

One of the easiest (and most often overlooked) ways to accomplish this is to utilize space in the figure where data could be plotted (which we call the dataspace), but no data exists. By moving the title of the figure, any explanatory text, and the labels that delineate individual series of data inside that dataspace, one is able to significantly increase the size of the overall figure, without losing information.

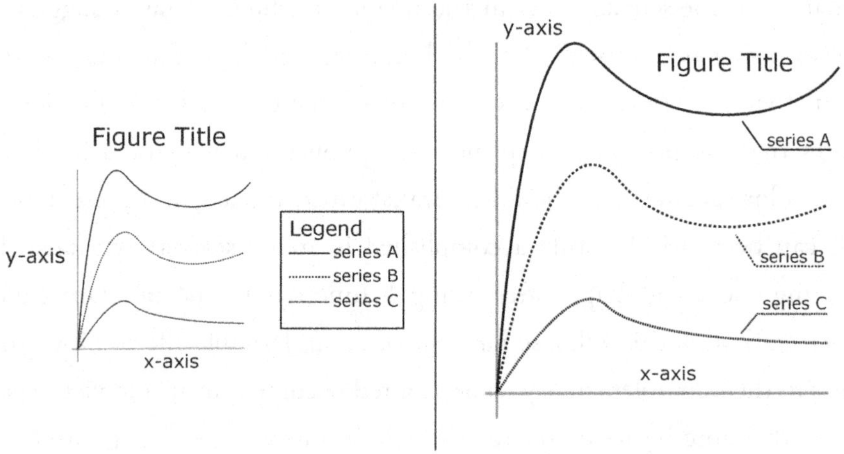

Figure 7-2: Two types of data figures. On the right, the dataspace is enlarged, and labels moved inside the axes, increasing legibility.

In the figure on the left, the dataspace is reduced in order to accommodate the legend, the exterior title and the exterior axis labels. In the figure on the right, by moving the title and the axis labels inside the dataspace, and removing the legend altogether in favor of labels directly on the series themselves, we are able to maximize the dataspace, which increases legibility, and makes it easier to distinguish one series from the next. One reason that figures like the one on the left have become more prevalent is that these are the types of figures that many data processing software packages produce. Images like the one of the right require more human input, but are vastly preferable, and can be easily achieved through one of a number of readily available software suites.

Using A Slide

Successfully delivering a talk supplemented by images is not just a matter of designing good images, but of using them effectively. As with slide design, there are a number of common practices in scientific talks that are less than ideal. As stated above, the guiding principle is using slides effectively is to remember that the slides are there to augment and enhance, not replace, the story being told by the speaker. To that end, there are five simple guidelines that will help you avoid the most common mistakes in effective use of images to enhance a presentation.

Rule #1. Face The Audience.

As discussed, this principle is the simplest, but it bears repeating, as it is so often ignored. Projection of voice, employment of body language, and effective posture all require that the speaker face the audience as she delivers her talk. One of the reasons that so many speakers fail to do this is that we want to be seen as impartial and dispassionate observers, viewing the figures like any other audience member, and some of us go so far as to move physically into the audience as we speak. This tendency is bad practice, and should be studiously avoided.

Rule #2. Use A Second Screen

One of the easiest and best ways to avoid violating Rule #1 is for the speaker to look at a screen other than the primary screen (the one the audience sees). This is almost always most easily done using the screen of the device (usually a laptop, desktop, or tablet) driving the presentation. All of the slide programs have a function called "presenter's display" or "presenter's view" that instead of showing just the slide, shows a thumbnail of the current slide, as well as the next slide, a timer, notes to the speaker, and more. These displays can be customized to suit individual preferences. Some modern programs even show these second screen displays on a handheld device (phone or tablet) that can also be used to control the slideshow. One of the devices should be used in conjunction with the primary screen to ensure that the speaker faces the audience.

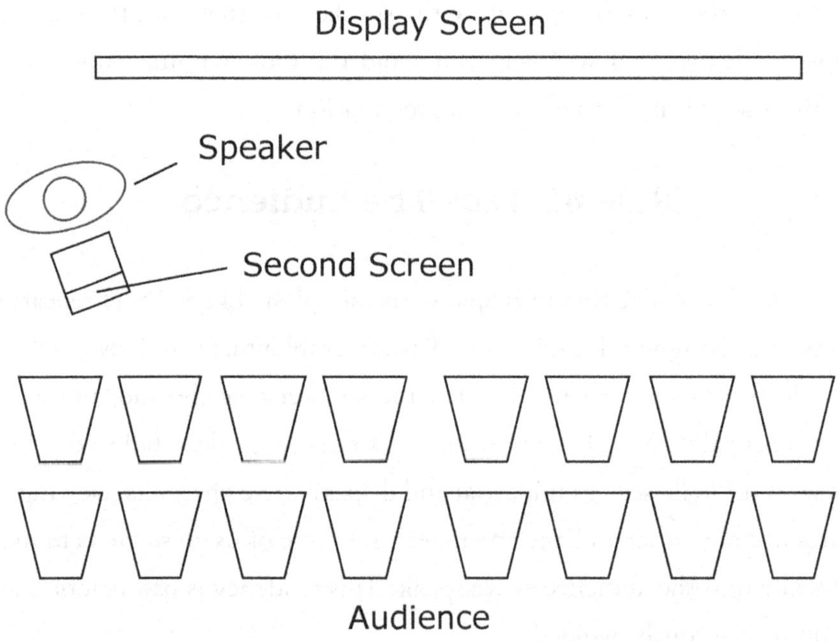

Figure 7-3: A typical configuration for a technical talk. The second screen should be used to show the speaker additional information.

Rule #3: Avoid Pointers

Except in a very small number of cases, a pointer is not necessary when speaking with slides. These devices are chronically over-used in scientific and technical talks. A well-designed slide will make the order of information apparent. While there will often be a section of an image of figure that you wish to draw attention to, you almost always know you wish to highlight that section when you are designing the slide, so it is far more effective to simply draw an arrow, label, or other callout into the slide itself, rather than gesturing. Using the pointer turns you away from the audience, makes you harder to see, and harder to hear. In those rare cases where a pointer is necessary, a physical pointer (or a yardstick, or your hand or finger) is preferable to a laser pointer, which is nearly impossible to hold steady.

Rule #4: Do Not Read Text Aloud

As discussed above, lines of text should be used sparingly, but when they are used, or if a full paragraph of text is deemed necessary, do not read the text to the audience. At a scientific or technical talk, you can usually assume your audience is literate, and if they are, they will immediately begin reading text you display on the screen, at a rate which will exceed the rate at which you can read it aloud. If you begin reading it to them, the two signals overlap one another and destructively interfere. The worst possible practice is to violate rules three and four simultaneously, reading a block of text to the audience, and bouncing a laser pointer along the text as you do so. This is not only immensely distracting, it is insulting to the audience's intelligence.

Rule #5: Keep The Focus on You

Because of the depth and complexity of many scientific and technical talks, it is easy to let the data or the figures being presented become the central feature, the thing around which the rest of the talk is built. But it is important to remember that the real focus must always be the *story*, the narrative that you are trying to construct. Information found in figures is only useful if the audience understands why they should be interested in it, and that understanding must come from the story being told by the storyteller. One simple technique for drawing the audience's attention back to yourself (and thus back to the story), is to punctuate a talk with a moment or two with no slide at all. This can be accomplished either by using a blank black slide, or (in most programs) by pressing the "B" key on the keyboard. This will momentarily mute the image, and shift the audience's focus back toward you.

It is also far easier to keep the audience focused on the speaker if you are able to move seamlessly in and out of the displayed image. Use of the black background and white foreground (discussed above) makes this infinitely easier, particularly in cases where the images is being generated by a front-facing projector. In these cases, a black slide allows the speaker to move into the image to gesture at a figure without distraction, while a white slide would cast light onto the speaker's face, blinding him to the audience, and casting a shadow on the screen behind him.

As slide software has proliferated, and slides have become more and more common, a practice has developed in some circles of intentionally loading slides with as much textual information as possible, in order to make them effective both as visual aids in a talk, and as documentation of that talk to be used later on. Such "slides" are not purely images, nor are they purely documents, but rather a hybrid of the two, and as such suffer from the compromises inherent in hybridized design. Anything which tries to be two things at once can never be very good at either of them. Countless attempts at the flying car or the road-going boat have died on the drawing board or in early prototypes. Communication products are no exception.

Hybrid slides may be serviceable in their dual roles, but they will never be outstanding. If time is of the absolute essence, a hybrid may serve, but it will always be far less effective than a purpose-built visual aid or a purpose-built document.

Perhaps the worst use of these hybrids is their employment in the classroom, where some instructors both show these "slides" and distribute copies of them to the students in lieu of asking them to take notes. This practice degrades the quality of instruction and dulls the wits of the pupils, and the world at large would be better off if it were abandoned.

8. Posters

```
                    MICHAEL
Come on, how often do you get the chance?
You can't just do it and not say it, come
on!
                     HENRY
Stop the presses!

                         -THE PAPER (1994)
```

One of the mainstays of scientific communication around the world, the technical poster can be a powerful and flexible format for sharing ongoing work. Individual technical posters provide brief "snapshots" of research, giving insight into a particular study or experiment, often when that work is still in its early stages. Poster sessions remain popular at scientific conferences around the world, because they offer attendees a chance to see several hundred of these snapshots in a very short period of time, and often provide the opportunity for people to meet and talk with one another in an informal setting.

A well-designed scientific poster therefore has a dual purpose. It is designed to serve as a stand-alone document, conveying an overview of the work and its importance; but also to serve as a conversation piece that fosters dialogue between the author and the viewer. In this chapter, we present the best practices for designing an effective technical poster. The chapter is organized into sections which address each step in the design process, presented in the order that they should be followed for best results.

A detailed discussion of how to handle the face-to-face interactions a great poster might initiate can be found in Chapter Twelve.

Writing

The first step in the process of creating an effective poster is to write down the ideas the poster will include. This is best done with the simplest possible tools, usually a pencil and paper. At this point, you need not worry about the specific language, the goal is simply to jot down the major points that you want to touch on. As always, the guiding principle should be selecting the overall narrative framework into which the information be placed, though in the case of the poster you will not have the space to explicitly state a narrative proposition. Instead, you should select the larger narrative you intend to use, and position the poster as one scene or episode from within that narrative. Decide who your someone is, and what something they want; the poster will catalog one of the troubles that character had or is having getting it.

Because the final format of this piece will be graphical, it can be useful at this stage to start thinking spatially. Writing a line or two on several small index cards or sticky notes will facilitate this, it allows you to easily rearrange particular sections. Generally, you will have a section explaining the context of the current work or the motivation behind it, two to five sections explaining the observation or experiment itself, and one summarizing the most important conclusions. Once you have sketched out these major sections, you can begin thinking more concretely about graphical layout.

Order

When choosing an overall layout for the poster, the most important thing to consider is the order of consumption in which you want or expect the viewer to read them. Because a poster must function as a standalone document, the reader or viewer should be able to digest the material without aid or facilitation. If you have two graphic panels which describe the current state of some system, and some future desired state, you want to present that material in a linear order. It is therefore necessary to provide visual cues which guide the viewer's eye, and avoid situations where the order of consumption is unclear.

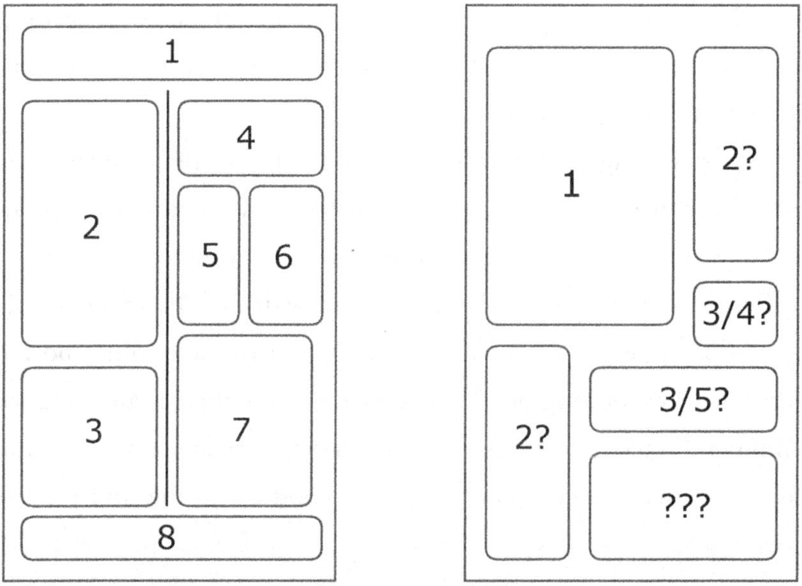

Figure 8-1: Poster layouts. The layout on the left guides the eye, using a simple line to demarcate two columns. In the layout on the right, there is no obvious order of consumption.

Choosing a layout will establish the overall graphical look of the poster. It is still advisable at this stage to work with simple tools like pencil and paper, as they afford the opportunity to quickly and efficiently sketch out and evaluate several possible designs before making a final selection.

Figures

Once a general layout for the poster has been chosen, you can begin assembling and modifying whatever accompanying graphs, photographs, or illustrations you wish to include. Since the poster will ultimately be printed with ink on paper, figures used in a poster need to be of a much higher resolution that those that would be used for a slide. This is particularly true of photographs, because as they scale up both the color and the clarity of the image will be distorted. Whenever possible, it is desirable to obtain figures (whether graphs or illustrations) as vector-based images, rather than raster (also known as bitmap) images.

A raster image is the most common type of digital image, and is the native format in which images are displayed by a computer. The raster image is essentially a large two-dimensional matrix of numbers. The dimensions of the matrix are the height and width of the image in pixels, and the value stored in each element of the matrix determines how light or dark the corresponding pixel will appear.[1] These images are common to all computers, since computers must ultimately display things as a matrix of pixels, and because computers are very good at storing and processing matrices.

Vector graphics, on the other hand, store the objects in the images as a series of mathematical expressions that represent primitive shapes. As such, they do not have a defined size in pixels. The great advantage of vector graphics is that because they store the underlying information that creates a figure, rather than storing a digital representation of that information, they can be scaled infinitely without loss of resolution, and can be easily modified or edited after scaling without creating artifacts.

---

1 For simplicity's sake, we are considering here a monochrome (black and white) image. In a color image, the elements in the matrix are themselves vectors of three or four elements, as discussed in the following section.

This scaling ability is critically important when choosing figures for a printed poster, as the resolution of the poster (usually expressed in dots per inch, or dpi) vastly exceeds the resolution of even the most advanced digital display devices. For this reason, the final file format sent to the printer will always be a vector-based file (most commonly the Portable Document Format, or .pdf). Creating that final file will be less error-prone if the constituent components are also vector images.

As with the creation of figures for slides, care should be taken to enlarge the dataspace, in order to make the figure as legible as possible. When creating figures for a poster, it may also be desirable to consider the color scheme employed in the figures themselves, in order to match those colors to the color palette chosen for the poster overall.

Color

The use of color can greatly enhance both the clarity and the visual appeal of a technical poster. Thanks to modern digital technology, a nearly limitless choice of color possibilities exist, but it is important to carefully and deliberately choose a palette of colors that will enhance, rather than detract from, the viewing experience. What's more, since the poster will be designed in software, but ultimately printed in ink, it is worthwhile to consider the differences in how color is perceived through those two technologies.

Colors are represented digitally as the sum of different proportions of the three primary colors, red, blue, and green. Together, these three primaries define a color space, or gamut, called the RGB space, which can be thought of as the spectrum of all possible colors. In the most commonly used color space, the primaries range in value from 0 to 255, so that each color displayed by the computer is represented by an array of three numbers within that range. The color red, for instance, is represented as: [255, 0, 0], with 100% of the possible red displayed, and both of the others set to zero.

Likewise blue is [0, 255, 0], black is [0, 0, 0], white is [255, 255, 255] and so on.[2] This list will sometimes include a fourth value, which is known as alpha, or transparency, with completely transparent set to zero, and completely opaque set to 255. This value is used when superimposing images one over the other to determine what percentage of the background shows through the foreground.

When choosing colors for a figure, or for the poster itself, it is important to note the specific numerical value of the color under consideration, for if you wish later to modify an element of the poster in another software suite, you will need to know the precise value of the colors in order to match them. Similarly, institutional logos or other branding will have specific standards for the colors they use, which can sometimes serve as inspiration for the selection of a color palette. Across computer devices and platforms, colors of the same numerical value will be identical, even if they appear slightly different, which is usually a result of differently calibrated display devices.

In the physical word, color is generated using an entirely different process. Color printing relies on a four-color process, known as the CMYK process, after the four colors of the inks used: Cyan, Magenta, Yellow, and blacK. These four components describe a separate color space, known as the CMYK space. Though they largely overlap, the CMYK and RGB gamuts are not identical, and as such there are some RGB colors which cannot be produced in CMYK printing, and vice versa.

In practice, the perception of color on the printed page can also differ wildly from the perception of color on a display monitor, because printed color can only be seen if light is reflected off of it from another source, and that light will have a color of its own. Though we sometimes think and speak of "white light," all light sources have a specific color profile. Midday sunlight is quite blue, while incandescent bulbs appear orange. The

2 These values are sometimes expressed in hexadecimal notation, where red would read: FF0000, blue 00FF00, white FFFFFF and so on.

fluorescent lights common in meeting rooms or conference halls often have a greenish tinge.

Because of these complicating factors, in practice the best solution is simply to test the various alternatives. A good rule of thumb is to select a small palette of three complementary colors that you will use in the poster. Print a test page that contains small patches of these colors, and look at them on the page under various lighting conditions. Once you have chosen a palette that you are happy with, write down the numerical values of the colors, so they can be used across software programs.

When choosing your colors, it is important to avoid colors that are very light, like bright yellows or other pastels, as they will not offer sufficient contrast against a white background, and to avoid choosing only reds and greens, as a significant percentage of the population has a form of color blindness that makes it impossible to distinguish between these two colors.

Fonts

Like color, modern computing offers a wide range of choices in terms of text font. The choices range from the traditional to the nearly illegible, and so it is usually best to make fairly conservative choices with regard to font. Broadly speaking, there are two major classes of fonts: serif fonts and sans serif fonts. Serifs are the small lines or "feet" at the bottom of letters. Fonts that have them are called serif fonts, those that do not are called sans serif fonts.

Figure 8-2: Serif and Sans Serif fonts.

Serif fonts are preferable for text on the printed page, as they significantly increase legibility (this book is typeset using a serif font), while sans serif fonts are commonly used for titles and headings. It is best to choose at most two different fonts that you will use in a poster, and stick to them. Avoid the temptation to switch between multiple fonts, as this creates a distracting sort of visual noise.

The proper size of text will depend on the setting in which you intend that text to be read. For a poster that will appear at a technical symposium or conference, the text should be legible from a distance of about two meters, which requires a font of roughly 32 pt. or larger. Note that the point in typography is a measure of physical distance,[3] so it is important to set up your document with the proper physical dimensions in order to make your text render at the proper size. The apparent size of a poster from roughly two meters is nearly identical to the apparent size of a standard sheet of office paper at a comfortable reading distance, so you can check size by printing a scaled down version of the poster on a single page. If you cannot read the text on the page, it will be illegible on the poster.

3 It's precise value is 1/72 of an inch, or 352.778 microns.

Software

Once you have written out the text for your posters, gathered your figures, chosen your layout, and selected your colors and fonts, it is time to begin constructing the poster in the software package of your choice. As discussed above, it is best to save this step for last, as working with simpler, analog tools like pencil and paper allow a freer breadth of experimentation. Of course the final product will be prepared as a digital file, and for that it is best to choose a software package that was designed with graphic layout and/or illustration in mind. These software programs offer several advantages, including the ability to manipulate images, the ability to control text wrapping and alignment, support for layering, bezier tools for drawing, and so on. Many excellent proprietary packages exist for those who can afford them, and a small number of open source tools are available (including the LaTex system), though these may have a steeper learning curve. Slideshare programs, while they may appear easy to use, are particularly bad choices for the design of technical posters, and should be avoided.

Once you have chosen your package, laying out the poster is straightforward, as you have done most of the work already. During layout, take care to respect the borders of the document, and be careful to make sure that the text and figures are precisely aligned, as small variations that may not be apparent on screen will be obvious when printed at full size.

Technology

Lastly, it is worth noting that while it remains useful, the technical poster (like many of the conventions of scientific communication), is firmly rooted in a largely outmoded technological model. Every scientific poster is created as a digital document, and nearly every audience member who comes to see them is carrying at least one digital display device. Yet poster sessions still rely exclusively on the printed page to convey information. Posters which contain embedded links like QR codes would offer viewers the chance to obtain contact information, links to published work, etc. Several tentative experiments with these technologies are now taking place, and it is possible they will eventually upset the current paradigm, allowing what we currently think of as a "poster" to encompass a broader range of digital media, including sound and video.

9. Motion Pictures

DENHAM
...In a few months, it'll be up in lights on Broadway: Kong, The Eighth Wonder of the World!

-KING KONG (1933)

The moving image is one of the most powerful forms of human communication ever devised. Since its introduction in the early 20th century, the medium, perhaps more than any other in history, has captivated the human imagination, and transformed the nature of human societies. Cinema, television, and video impact modern culture like almost nothing else, shaping global opinion on everything from fashion to politics.

Thanks to modern technologies, there is more video content than ever before. The production and distribution of moving pictures once required infrastructure on an industrial scale, but those tools are now readily available to the average consumer. Most modern smartphones are reasonably capable of capturing video, and in many cases capable of editing and distributing it. Web-based distribution platforms enable anyone to become a video producer, and business models are rapidly evolving that allow those small producers to sell advertising, collect revenue, track audiences, and much more.

Yet despite these changes, the fundamental principles of film and video production remain the same. In this chapter, our goal is to highlight some of the most important of these principles, from developing a concept to distributing a finished product. Our treatment of these topics is necessarily a broad overview, and is intended for the complete amateur with no prior training or experience in film or video production. The reader interested in going further should consult one of the many texts that deal with these subjects more formally and in greater detail.[1]

Pre-Production

Perhaps the most important part of the video production process is the work that happens before there are any lights, cameras, or action involved. The process of making a video or film is complex and intensive, involving hundreds of separate steps. It is therefore essential that before you embark on that process, you have a very clear idea of precisely what kind of video you intend to make, and how you intend to make it. In the film business, this process is referred to as pre-production, the work that is done before capturing any footage.

As discussed in Chapter Three, this process should begin with careful consideration of your intended impact and your intended audience. The world of cinema and television is one where an almost limitless expression is possible, but the medium is subject to a long list of constraints, including both technical standards and artistic conventions. Online video is currently less technically constrained, but artistic conventions are rapidly emerging. In all cases, it is important to choose a platform which is able to reach the audience you hope to impact. Are you making a five-minute web-based video, or a two hour documentary film? The types of audiences who will watch each of those things are different, and it is important to choose the format that best suits your audience.

Likewise, it is essential to decide at the earliest stage what story you

---

1 One of the very best is *The Lean Forward Moment*, by Norman Hollyn.

hope to tell. What will the narrative proposition be? Even in a thirty-second television advertisement, there is almost always a clear narrative. Someone wants something badly and has trouble getting it. Decide what narrative you will use, and how you will go about rendering that narrative visually. Even if you are not certain precisely how the narrative will unfold (e.g. if your film will largely be based on interviews you have not yet conducted) you should have a good idea of what the narrative framework will be, even if you do not yet know how it will ultimately be resolved.

Once you have a clear sense of where your piece will be viewed, and what story you plan to tell, you should begin making an outline of the separate visual and auditory elements that will make up the finished film. There are many possible ways to go about this, but the most common is with a screenplay, a written document that describes what we will see on screen, and what (if anything) the people we see will say. Screenplays used in films and television are specialized documents that obey a long list of historical conventions in terms of the way they are written and formatted. It is not necessary for the beginner to understand or obey any of these conventions, but it is necessary to have some documentation which outlines what will transpire on screen.

When preparing your screenplay, it is important to follow what many consider the first rule of visual storytelling, neatly summarized as:

Show. Don't Tell.

This principle is important because the power of film comes from its ability to allow the audience to observe the evolution of a story directly, rather than hear about it from the mouth of storyteller. It allows for the transmission of ideas and feelings without the adulterating interference of language.[2] One of

2 Cinema thrived for nearly three decades before dialogue was technically possible. These "silent" films were not only compelling, but universal, as they could be understood by audiences around the world, regardless of their native languages.

the great powers of actors onscreen is that they can show us their emotions, allowing us to feel as they do, without needing to express it in words. Embrace this principle as early and fully as possible, when you are planning what you hope to capture. Rather than hearing someone say: "It was a rainy day in London…", show us the rain puddling in the streets. Don't feature an astronomer saying: "The telescope is larger than a city bus," show us the astronomer standing next to the telescope. Plan from the beginning to tell a visual story, a story told with pictures more than with words. Even if your piece will ultimately have lots of words (through voiceover narration, or on-camera interviews, etc.) make sure the story is told mostly in pictures. Show. Don't tell.

One of the tools that filmmakers use to assist them in this process is storyboarding, the process of creating visual sketches that outline each of the individual shots that the finished scene will comprise. These sketches (known as the storyboards) are particularly used in programs that will make extensive use of visual effects, because those effects are costly and time-consuming to produce. Storyboards offer a visual representation of each shot, and allow you to visualize how those shots will work together. While they may not be necessary or appropriate for every production, they can be useful for more complex sequences, and many modern digital tools exist to assist in their creation.

Whether you have a fully fleshed out screenplay and a complete set of storyboards, or just a series of index cards with notes of what shot you want to capture, it is essential that you have some plan before you begin. Once that plan is in place, you are ready to reach for the camera.

Picture

Once highly technical and very expensive, motion picture capture is now a nearly ubiquitous technology. Almost every modern smartphone is equipped a rudimentary capability to capture video, and almost all high-end cinema cameras are now digital. In between those two extremes, there are a dizzying array of options, offering choices to suit almost every style and any budget. Here, we look at a few of the most important aspects of cinematography, intended to give the complete beginner a basic foundation in the art form. The interested reader should consult one of the many available resources for more information.[3]

The first important consideration when capturing video is called framing; the choice of orientation of the capture device. The dimensions of cinema frames vary somewhat, but almost all are rectangular, and are always presented in a horizontal, or landscape orientation. This is because this orientation mirrors the field of view of human vision, since our eyes are arranged horizontally, not vertically. If you are shooting with a device whose orientation can be changed (like a smartphone) it is far preferable to hold the device sideways, in order to capture images in a horizontal (landscape) orientation.

3 Several excellent tutorials can be found online at the Vimeo Video School.

Figure 9-1: Film or video should always be captured in the "landscape orientation, with the long axis of the frame parallel to the horizon.

Figure 9-2: Composition using the "Rule of Thirds." This rule can be used as a guideline to almost any shot.

Given this horizontal frame, the next important consideration is called composition, the arrangement of the elements of the picture within the frame. Given a certain subject (e.g. a person being interviewed), there are a wide range of possible ways to place that subject in the frame.

A broad consensus exists within the cinematography community that one of the most aesthetically pleasing compositions is one which obeys the so-called "rule of thirds." The idea behind this convention is that the frame can be divided up by lines that demarcate three equal sections along both its height and its width. When framing the shot, the goal is to align the major subject of interest (the eyes of the person, the horizon of the landscape, etc.) with one of these lines. While this convention (like all artistic conventions) can be broken, it is a simple guideline that can help create shots with a more professional look.

A crucial component to consider when choosing the composition of your shot is how the elements in the scene (most importantly, the camera) will move. For a variety of reasons, images where the camera is moving quickly or erratically are at best distracting, and at worst can leave the viewer feeling queasy. In simple shots like interview footage, it is usually best to keep the camera immobilized, either on a tripod or other device.

There are cases where it is preferable to have the camera move, either to follow a subject, or show a broader subject than can fit into the frame. In these cases, care should be taken to minimize unwanted movement of the camera. In professional cinema, a menagerie of mechanical equipment is employed to achieve this, but the amateur can also obtain good results without specialized gear. The basic principle is to minimize vibration and oscillation, and to limit the degrees of freedom of movement of the camera. A simple device like a monopod can assist in this effort, as can any mounting device which makes the camera heavier. The general rule should be to make movements slow and deliberate, and moving shots usually both start and end at rest. Because of the rapid proliferation of small cameras that can be mounted almost anywhere, audiences have grown more tolerant to the rapid, often shaky footage that they produce, but smooth, fluid, deliberate shots will still look and feel more professional.

The final (but probably most important) thing to consider when setting up a shot is the lighting. Professional crews spend an enormous amount of time and energy arranging and adjusting lights to obtain a very specific look and feel, but simple lighting principles can be used by any amateur to improve the appearance of most shots. Lighting is particularly important when using small handheld systems, as their small lenses and sensors usually result in very poor performance in low lighting conditions. When working with these systems, it is worthwhile to find and use the best lighting at your disposal.

The most important element in most lighting setups is what is called the "key" light, the main light source that illuminates the subject. In many situations, your key lighting will come from the sun, either outdoors or from a window in a room. In either case, it is best to arrange your shot so that the camera is placed between the subject and the key light. This way, the camera is looking at the subject nicely illuminated by the light reflected from the key, rather than looking into the light itself. This is especially important in systems where the exposure is adjusted automatically, as the camera will adjust to correctly expose the brightest thing in the frame, which will leave the subject too dark. [see figure]. The key light is typically placed so that it shines on the subject from an angle, which creates shadows that give definition and a sense of depth.

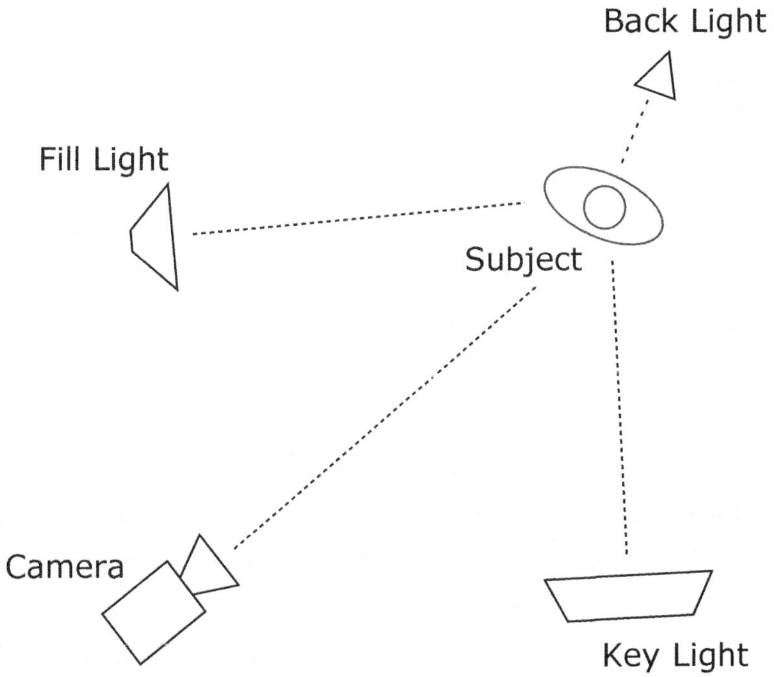

Figure 9-3: Basic three point lighting setup, one of the most common lighting configurations.

In addition to the key light, most lighting setups also include what is called a "fill" light, which typically illuminates the subject from the side not illuminated by the key light, in order to brighten the shadows created by the key light. A third light, known as the "backlight" is also often used to light the subject from behind, which separates the subject from the background. These three lights together form the most common lighting setup, known as three-point lighting.

While film and video professional have specialized equipment for lighting these situations, an amatuer can utilize these same principles to create lighting that has some of the same visual appeal. The key light may be light from an open window, while the fill and back lights are simple household lamps; used together they will still give more definition to the subject, and a sense of depth to the image, than would otherwise be possible.

Sound

Perhaps the most often overlooked element in film and video production, it is difficult to overstate the importance of high quality audio. Many amateurs who have been extremely careful and conscientious with capturing the image spend little or no time thinking about sound, and even more than picture, the quality of the sound recording and editing is what makes the difference between a product that looks and feels professional and one that has the feel of something clunky and homemade.

There are, broadly speaking, two types of sound that are used in film and video production: production sound and post-production sound. Production sound is the sound that is captured on-set, simultaneously with the picture. This sound (e.g. the dialogue captured from an interview subject) is synchronized with the picture, and thus sometimes called "sync sound." Post-production sound is everything else; the soundtrack, sound effects, and any background sounds added after the fact.

As with almost everything else in video production, the key to achieving good results in sound recording and sound editing is to have a plan before you begin. All video content will have sound of some sort, but not all videos require production sound. For example, your video may consist of voiceover narration and a montage of footage of field observations made of some species in some remote location. These films often feature sound elements, (the rush of wind, the sound of a babbling brook, the call of the species in question), but in practice those elements are almost always added during the editing process, rather than captured with the footage. Decide before you begin shooting if you are capturing sound and video, or just video.[4] If you have no intention of using the sound, you are then free to

4 Footage captured without sound is commonly referred to as "MOS," which

do several things, like direct the action on camera, or make audio notes to yourself about what you are seeing.

In many cases, you will want synchronized sound, the most common of which is a subject speaking on screen. In this case it is important to do all you can to capture the best sound possible. In contrast to video technology, the technology for capturing sound has changed relatively little in the last several decades. Furthermore, while devices like smartphones have very good capabilities for capturing video, their capacity for sounds capture is not nearly as good. Far better results will be obtained by using specialized sound equipment. The most common types of microphones are lavalier (commonly called "lav") mics, which clip to the subject's apparel, or directional "shotgun" mics, which can be targeted to a specific source. Depending on your equipment, these microphones can either send their signal directly into the camera, or can capture sound to an independent device like a mobile recorder, an arrangement known as "dual system" sound. If no other options exist, one can create a dual system using two smartphones, one in the breast pocket of the subject recording audio, and another held by the operator a few feet away recording video. Regardless of what equipment you have available, it is important to place the microphone as near the subject as possible, and to minimize background noise in the space. When recording interviews, always capture 15-30 seconds of silence, known as "room tone." This piece of background sound will be incredibly useful in sound editing, discussed below.

In many cases, sound can be captured without any picture at all that will be useful in the final edit. The sound of a machine in action, or the rumble of a truck down a rough road, the sound of a jet engine spooling up, or of a wave crashing on a beach may all be sounds that you intend to use in your finished piece, but you don't often need to capture those

---

according to legend is an acronym for the broken English "Mit Out Sound," as spoken by an early German filmmaker.

sounds simultaneously with the picture. Spend time thinking carefully about sound from the beginning of your production, and invest in the best sound recording equipment you can afford. That investment will pay huge dividends later on.

Sound recording should not be an afterthought, as it is an essential component of a finished product. Without knowing it, audiences judge the quality of a production based on the sound, even more than the picture. A video with sub-standard picture but great sound will feel more professional than one with exceptional picture and terrible sound.

Editing

Once you have captured your footage and your sounds, it is now time to put all of these elements together into a finished product. For historical reasons, this process is referred to as editing, though it bears little relation to the process of revision and error-checking in writing that goes by the same name. In film the editing process is where the piece is created; what we call film editing is more accurately thought of as film composition.

Over the last century, as the art of film has evolved, a very specialized language of editing has emerged. Because of film (and television's) ubiquity, this language is spoken and understood around the world, despite the fact that at first glance it makes little logical sense.

Consider a scene from a typical drama, in which a man and a woman sit across a table talking to one another. The scene likely contains several cuts, places where one shot ends and another begins. The first shot may be from across the room, where we see both people seated. The next shot might be over the man's shoulder, looking at the woman as she speaks. We then cut immediately to the reverse angle, a shot looking over the woman's shoulder as the man speaks.

This arrangement of images seems nonsensical, as the camera is effectively instantaneously jumping from one point in space to another. The sequence shows us something that we never see in reality. Yet the audience accepts it (indeed, may not even know it is happening), because it is part of the language of cinema that we have all come to know. The conventions of that language are specialized and complex, and students of film (like students of most arts) spend most of their educations mastering those rules, if only so that they can break them effectively in pursuit of their own art. For our purposes here, it is enough to know that editing is where films and video are truly made, and that a full understanding of the art form must be based on an understanding of edits.

For the amateur, the process of editing will largely be one of assembly, putting together the pieces of picture and sound that you captured to tell that the story that you intended in your outline or screenplay. In practice, this will almost always be done with the help of a video editing software package, of which myriad variations exist. Though these software packages vary immensely in cost and complexity, the majority share a short list of basic features, and nearly all are based on a simple theory known as three point editing.

The process of video editing is one of taking segments of video (often known as "clips") and pasting them together one after the other to create a finished video. This is accomplished within most editors by first selecting the segment, and then placing that segment onto a timeline. This can accomplished in one of several ways. There are "in" and "out" points on the source footage, the places where the segment begins and ends, and there are "in" and "out" points on the timeline, the points in the finished video where the shots appear. Given these four points, one can choose any three points, and that will determine both the length and location of the shot.[5]

Audio editing works precisely the same way, though the degree of

5 One cannot choose all four points, as that would overconstrain the problem, e.g. trying to paste a three-second clip into a ten-second space.

independent control you have depends on the particular editing software. One thing particularly to avoid in audio is what's called a "jump" cut, meaning a place where the background audio abruptly changes. A good example of this is a sequence where we see an interview subject talking about some object, then see a photograph of the object, then see the subject speaking again. If we hear no audio when we see the photograph, there will be jarring and obvious cuts in the sound before and after we see the photograph. Inserting a few seconds of room tone under the picture will create a seamless and fluid sequence.

Because of the complexity of video editing, the best advice for the complete amateur is simply to experiment. Choose an editing suite, and make something. It can be simple and short, as little as 30 seconds, but plan it, shoot it, and then cut it together. This is by far the best way to learn both the tools and the techniques. Start with very simple elements, and as you gain confidence, start exploring your options, like titles, effects, transitions, etc. The general rule during these experiments is do not add new tools to your repertoire (or move to a more complex software system) until you feel you have mastered and exhausted the creative possibilities of the tools you are already using. A simple tool in the steady hands of a confident user will yield better results than a complex tool wielded by a skittish amateur.

Distribution

Once the editing process in complete, it is time to share your creation with the world. This process is much simpler technically than it once was, thanks to digital technologies, but with the ease of use comes further complications. One can publish one's video content with just a few clicks, and make it available worldwide, but just because it is available doesn't mean anyone will actually see it. Digital video sharing sites are invaluable resources, but because they have lowered or eliminated the barriers to widely sharing video, the average user is awash in a flood of content.

The best choice for the distribution of video content will depend to a great degree on what your intended impact is, and thus who your audience is going to be. Different video sharing services attract different audiences. While you might choose to post your video in more than one location, you will have to choose how you allocate resources to draw attention to it.

The process of posting your video to a video sharing service is an easy one, but it is not the end to the question of distributing your content. The real challenge is then to direct traffic toward that content. Drawing attention to your work is one of the hardest parts of the process, but it is no different than drawing attention to anything else, and in the 21st century, it is most likely that you'll be using the tools of social media, discussed in detail in Chapter Eleven.

10. News

KANE
I run a couple of newspapers. What do you do?

-CITIZEN KANE (1941)

In modern democracies, there is no more influential institution than the free press. The fourth estate has been one of the pillars of western societies for at least the last three centuries, playing a role in all our affairs, and science is no exception. In this century, the technologies, techniques, and business models upon which the news media rely are all changing, but the fundamental principles of journalism remain.

Science journalism, more than other segments of that industry, has suffered in recent decades. Major newspapers have scaled back or eliminated their science sections, and many of the magazines devoted to science have disappeared. Science television programs have become harder to find, and science writing is becoming a more and more difficult way to make a living. At the same time, science itself is exploding, creating new knowledge and new applications at a faster pace than ever before. Thus there are more and more stories competing for less and less space in the news. In this environment, it is more important than ever for scientists themselves to have a "news sense", to know what is and is not newsworthy, and to craft their communications in the clearest and most concise way possible, to increase the likelihood of news coverage.

In this chapter, we introduce the basic concepts of newsmaking in science. We introduce the basic theory of journalism, the most common structure of a news story, and then discuss preparing press releases, handling interviews, and writing opinion pieces. Our hope is to give the reader a general overview of the process by which scientific findings make their way into the news. A fuller treatment can be found in one of the many volumes devoted to the subject.[1]

Theory

The word "news" in the English language has a very simple etymology, which does much to help define what news is. The word emerged in the 14th century as a plural of the word "new." This change is not difficult to imagine, as it requires merely that citizens in the village square, asking one another: "What is new?", eventually began to phrase the question: "What are the news?", which then became: "What news?" (as it appears in Shakespeare). This origin clearly outlines what one can consider the first theorem of news:

If it isn't new, it isn't News.

This simple idea is the guiding principle one must keep in mind when dealing with journalists. Their job is to gather, process, and disseminate the news. Scientists, given the chance, are often tempted to try and use the news media as a platform to disseminate information, thinking that their moment in front of a camera or microphone is an opportunity to *teach.* Not so. The nature of the business is such that it concerns itself only with reporting that which is new or different, not with disseminating long-established facts. As

1 The best by far is *Am I Making Myself Clear?* by Cornelia Dean, a former science writer and editor at the New York Times.

such, when dealing with journalists, it is important to constrain yourself to that facet of your work which is new, the thing you did that is different than what anyone else has done before. If it isn't new, it isn't news.

News in its modern form emerged during the 19th century, thanks in large part to the rise of the newspaper. Though the number of newspapers printed around the world has plummeted in the last several decades, the newspaper remains one of the mainstays of news media, and many of the conventions of the newspaper story are used on other platforms.

Newspaper stories are almost always structured using what is often called the inverted pyramid. In this form, the broadest possible summary, an overview of the facts and their importance, is presented at the very beginning, or top of the story. This introductory section is known as the lede, and it summarizes the main points of the story in as little space as possible. The lede focuses on the facts: the who, what, where, and when of the story. The clearest examples of ledes can be found in news stories that consist mainly of straight reporting of events, such as coverage of disasters. A typical lede in such a story might read as follows:

A fire broke out on Main Street yesterday afternoon, destroying two buildings and causing dozens of others to be evacuated.

As the story progresses, its focus narrows, exploring the details behind the lede. The subsequent paragraphs explain the background of the situation, how things came to be as they are, the precedent that led up to events, etc. In the example above, it might tell us what investigators suspect caused the fire, how quickly it spread, whether anyone was hurt, etc. The following paragraphs will either give more background information (e.g. this is one in a series of fires), address likely consequences (e.g. city leaders are calling on the mayor to tighten regulations to improve fire safety) or adding detail that paint a richer picture of the event (e.g. "The smoke choked out the sun," said one survivor).

This form of story emerged in newspapers for the simple reason that publishers were well aware of the fact that not every story would interest every reader. The stories compete with one another for a reader's attention, and thus the writer's job is to give the broadest possible view at the very beginning, so that the reader may decide for herself whether to continue reading. The structure conveys the maximum amount of information in the minimum of time and space, and it is used throughout journalism, in print as well as in television and on radio.

It is worth noting that this structure is almost precisely the inverse of the manner in which one would present information in a scholarly publication. [See Figure]. When writing in the sciences, we nearly always begin with the background information, establish the current state of the field, define the question, describe our methods, present results, and only in the end offer our interpretation of what we found and what it means. That structure will not work in journalism. We cannot write for the news the way we would write for a scholarly publication.

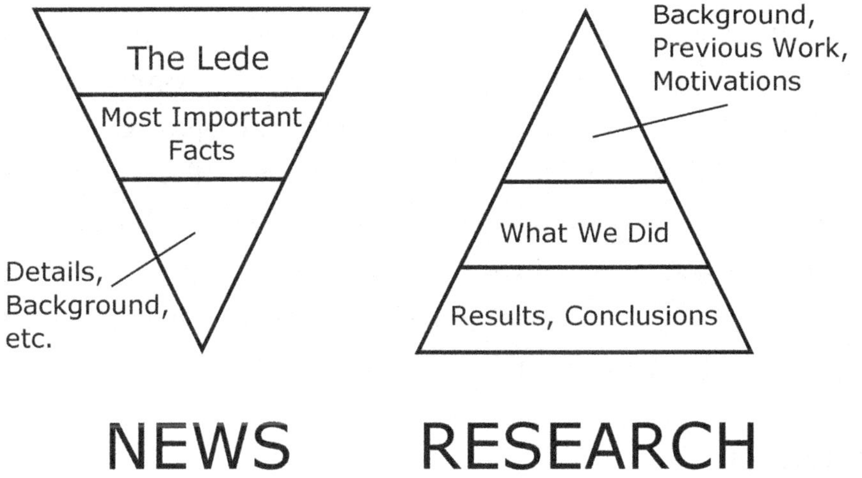

Figure 10-1: Structure of news vs. structure of scholarship.

There is another important difference between scholarly writing and news writing, and it involves parts of speech. Journalists are constantly thinking about, looking for, and employing verbs. Verbs are active words, words that convey something taking place, something dynamic. The verb, writers often say, is what's *happening*. Verbs are able to describe a much more vivid, dynamic, and thus engaging scene than nouns and adjectives alone. In theater and cinema as well, actors and directors are also always thinking about and talking about verbs, describing what one character does (or tries to do) to another character by finding that character's verb, because the verb encapsulates the action.

In the sciences, we make much more limited use of verbs, focusing instead on nouns. Even in cases where we are describing active language, we tend to use the noun form of verbs. Scientists desribe excitations, or mutations, they rarely say something excites or mutates. This contrast between scientific writing and journalistic writing can be seen clearly by looking at two publications side by side. Leaf through a copy of a recent scientific journal, reading only the titles of the articles. Look for verbs. You won't find many. The titles tend to be only nouns and adjectives, the names of the thing examined or explored. Now leaf through a newspaper, reading only the headlines. Look for verbs. Almost every newspaper headline contains one, because the verb tells the reader what's happening.

This dichotomy is important to remember when framing a research result for dissemination through the media. Journalists need to be able to describe your work using verbs, so you can help them by providing some; if you can talk about your work using language that is not just descriptive, but active, you will be far ahead of the competition.

Practice

In practical terms, the process of publicizing and disseminating research results through the media begins when some significant result has been achieved. If you have done something new, there is a chance that it will be news. For scientists working at institutions, the first step in the process is to contact your institution's public information officer, or PIO. Sometimes called press officers or public relations officers, these professionals are people who are tasked with helping translate your research work into a form that will be most readily accessible to journalists. Many PIOs are former journalists themselves, or have contacts within the media that can greatly amplify your message.[2]

PIOs can only help you, however, if they are aware of your work in advance of the news event. Commonly in science, that event is the publication or release of a major finding in a scholarly publication. In that case, the proper time to contact your PIO is when the paper has been accepted by the journal, and a publication date has been set. This gives sufficient time for the PIO to read and digest the material, develop a strategy, etc. Whether the

2 A full treatment of the roles and responsibilities of the PIO can be found in *Handbook for Science Public Information Officers*, by Matt Shipman.

event you seek to publicize is a paper, or a conference, or the opening of a facility, it is essential that you contact your media relations office as early as possible.

Because PIOs at large institutions have so many demands placed on their time, and because many institutions lack the resources the adequately staff a full office of PIOs, it is important for you, the scientist, to understand the process by which findings are prepared for media dissemination. The details of that process are shifting with modern media business models, but the traditional process remains the most common one, and it begins with the preparation of a press release.

A press release (also called a press notice, a news release, or a media release) is any document prepared for and sent to journalists. The purpose of the document is to make writers and editors aware of the news described in the release, in the hope that they will use it as the basis for a news story. A traditional press release is not a news story; rather it contains the information that one would need to construct a news story. The hope is that a reporter, upon receiving the release, would do more reporting on her own, interviewing the relevant players, seeking comment from outside sources, etc., before writing the final piece.

When preparing a press release based on a scientific finding, such as the publication of a scholarly article or the announcement of a major finding at a conference, the main task is to distill the main ideas. It will never be possible in the space of a news article to fully convey all of the relevant science. The goal instead is to identify the most important points, and highlight them in the release. A typical release is short, comprising three to five paragraphs, and no more than 500 words.

The first step in this process is to identify what is new. Your work has shown something that hasn't been shown before. It's the first of its kind in some way. Identify that attribute. The aspect of your work that makes it "novel" is the thing that is most likely to make it newsworthy. Try to craft a simple sentence that describes what that new work is, using a verb. That sentence will be the basis of the lede of your news release.

The second step in crafting a release is to focus on impact, and the best way to do that is to frame the impact as a component of a narrative. There is something important about your work, a reason why people care about it. Identify that thing, and frame it as either the resolution to, or a step toward, some narrative proposition. As we discussed in Chapter Five, there are several possible narratives to choose from. Choose one, and use it to frame the importance of your work. Someone wanted something badly and has had trouble getting it. Your work usually has something to do with the trouble. Describe it in those terms. The impact should be mentioned in the opening paragraph of the release, and can be expanded upon in a following paragraph.

The third step in the process is to identify the next logical step that your work suggests. Science is incremental by nature; every answer poses new questions. Briefly identify what that next step is, and what you you hope to achieve through it. If possible, make that next step part of the narrative that you identified when describing your impact. You have just conquered a piece of the trouble that someone was having getting something. Tell us what the next step is that will bring us closer to that goal.

These three steps should constitute the basis of the press release. Describe what you've done (and what's new about it), explain why it is important (and who should care), and then identify your next challenge. It can be helpful to think of these three steps as the answer to three simple questions:

What?

So What?

Now What?

A good press release will answer all three of these questions, in crisp, active language that uses verbs.

The last technique that is often employed when constructing a release is to find what is called a "hook," or a "news peg," something about your work that is related to some other story currently in the news. This might be anything from relating your work to some topical issue in global geopolitics to referencing a popular film playing in the cinema. The point of a hook is simply to grab the reader's attention with something that makes your story stand out among the parade of news stories about science that pass daily before our eyes. A good public information officer can be invaluable in helping you identify and employ a good hook, and one of the reasons it is essential to enlist your PIO's assistance as early in the process as possible is that it allows him or her to be on the lookout for a good hook to your story.

In practice, thanks to digital media, the traditional process of writing and sending a release only to reporters is becoming less common. Many web services simply re-print and run press releases verbatim, and more and more institutional news offices choose to publish stories on their own news services. As a result, the line between a traditional press release and a news article has begun to blur, with more and more documents taking a hybrid approach that bridges the gap between the two. Yet a good story should answer the same questions that a good release answers, so if a scientist can think carefully and effectively about those answers, she will be more effective dealing with both press officers and with members of the press.

Interviews

One of the best things that can happen to a researcher as the result of publicizing a story is to receive a request for an interview. Interviews are a sign either that a writer is interested in your work, or the hosts of a show are interested in having you as a guest. Interviews offer scientists the opportunity for a much broader platform than we are typically afforded, but it is important to approach them carefully and deliberately, lest we squander an opportunity to be heard.

Interviews generally come in two forms. The first is an interview request that stems from a press release or news story, the second is a request to provide commentary or expertise about a story happening somewhere else in the news. For the sake of clarity, we will call the former an inside story, and the latter an outside story. The tactics employed in preparing for and executing a media interview are mostly the same in both cases, with a few exceptions.

The first and most important thing to do when you receive a request for an interview is to **reply immediately**. Journalism is a fast-paced business, and if a reporter does not hear back from you, they will simply move on to another story or source. It is thus essential to respond quickly. Whenever possible, use the telephone rather than relying on email or other text-based communication. Most reporters prefer voice communication, as it is more immediate, and conveys more emotional content. When responding to reporters, the first things you'll need are the answers to a few simple questions.

First, find out who the interviewer is and who they work for. If you have never heard of the publication or network, do some research. Try to find out if they are a reputable news organization, as some organizations masquerade as news services but in fact have some other agenda. How broad is their audience? Are they local, regional, national, international? Knowing who their news is likely to reach will inform what you choose to say.

Next, find out the format. Is this an interview for a written piece, as in a newspaper or website, or is it a recorded interview for use on television or radio? If it's an inside story, chances are they are looking for a good quote from a researcher, whereas in an outside story they will usually want either background information or an informed opinion. In either case, and despite the format, you will perform far better in an interview if you have taken the time to prepare what are called talking points.

Talking points are brief prepared answers to the most common questions you are likely to face in an interview. Reporters will ask the same kinds of what/so what/now what questions that you addressed in your press release, so use that release as the basis of developing your talking points. When preparing talking points, it helps immensely if you take the time to write them down. Space in news is at a premium, so good writing requires thrift. It is far easier to edit on paper (or onscreen) than it is in your head. By writing out your talking points, you will be more able to see places where you can omit needless words, employ a more active voice, etc.[3]

Once you have crafted your talking points, take the time to memorize them, verbatim. This exercise serves two purposes. First, it arms you with prepared answers, so that you know you have something of value to say. This helps put the mind at ease, and is particularly useful in the case of a live interview, where most people are struck with a great deal of nervous energy. Having memorized prepared statements will calm and slow the interviewee. Second, having a memorized talking point allows you to pivot from the question that was asked to the answer you wanted. For example, if the interviewer asks a question about a field outside of the scope of your work, you can say: "That's a fascinating area being addressed by some of my colleagues, but what I'm excited about is..." And then proceed to your

3 All of this advice and much more can be found in the slender tome that many writers consider the best book about writing: *Elements of Style*, by William Strunk and E.B. White. The book is colloquially known as "Strunk & White."

talking point.

When being interviewed by a print journalist, it is perfectly acceptable to tell them you need more time to formulate an answer and get back to them. This is particularly true when dealing with outside stories, where you may be asked to comment on material that you are hearing about for the first time. Ask for thirty minutes to think it over, then formulate an answer and get back to them.

If the interview is being recorded for radio or television, don't be afraid to stop yourself and start again if you fumble over a word or phrase, or give an answer to a question that you think is overly verbose or unintelligible. The interview will be edited before it is broadcast, and those stumbles can be removed. Simply ask the interviewer if you can start again, take a deep breath, and try answering the question again. Answers often sound clearer and more concise on the second try.

If you are being interviewed live on television or radio, obviously you have to keep going, which is why it is doubly important to communicate beforehand with the interviewer, so that you know the approximate length of the interview, the type of questions you are likely to face, etc. Memorize your talking points and try to relax. Live media interviews are probably the most difficult skill to master, so don't be discouraged if you don't perform well. Do your best, and then think about how you can improve next time.

When being interviewed, particularly for an outside story, you will have far more success if you are willing to express an opinion. As scientists, we tend to naturally exclude our own thoughts or opinions on things, sticking instead to what we can show based on the data. While appropriate in a professional context, this approach does not work well in the media. The readers or listeners don't have the time or the training to parse the data; they want to hear the opinion of an expert. Researchers who are willing to express their opinions, and are able to do so in a clear and concise language, are much more likely to receive further requests for interviews.

Lastly, it is important to remember when speaking to journalists that everything you say is "on the record." Unless you are speaking to a journalist who happens to be a trusted personal friend and has assured you she will preserve your anonymity, assume that anything and everything you say may appear in print. Thus it is essential not to say anything that you don't want published. You can go a long way to ensuring this by crafting and using good talking points, but when speaking off the cuff, remember that what you say may be broadcast to the world.

Opinion Pieces

We close our discussion of science in the news by exploring what is perhaps the media channel least utilized by science: the opinion piece. Opinion pieces can take several forms, including letters to the editor, op-ed pieces, editorials, commentaries, etc. These pieces run in every major newspaper, in many magazines, and in publications ranging from the reports of professional societies to university alumni magazines. They are short, compact expressions of a particular point of view, and offer a unique opportunity for scientists to share their voices.

That said, opinion pieces are not easy to write. Presenting a clear, cogent, consistent argument in as little as 100 words is no mean feat. The challenge seems doubly daunting for scientists, who too often fear stating their own opinions. This reticence may stem from the concern that it will compromise our professional reputations, that if we are seen in the press stating our own views as if they were fact, we risk the ridicule of our colleagues. Yet these fears are overblown, because opinion pieces are just that: opinion pieces. The reader knows that the author is presenting their opinion on the matter *a priori*.

The best (perhaps the only) way to learn to write these pieces well is to study, and to practice. Read the letters to the editor. Read the op-ed pages. Read the commentaries in your favorite magazine or journal. Pick out the pieces that you like most. What do they have in common? Chances are, the pieces that stand out are the ones that have a clear *voice.* One of the keys to writing these pieces well is to develop a voice that works for you.

When you read or see something in the news that you disagree with, write a short statement articulating your disagreement. Then edit it. Show it to someone who you think is a good writer, and who doesn't work in your technical specialty. Ask for feedback on how you can make the argument clearer. Then follow that story over the coming days and weeks. Chances are someone else disagreed with it too, and wrote about it somewhere. Read those responses, and compare them to your own. Over time you will develop a sense of the things you want to speak about, and a voice with which to speak about them.

Opinion writing is a complex skill that takes years to develop, but society would be better off if more scientists took the time to master it, and engaged more often in public discourse. The need for public engagement by scientists is greater in the twenty-first century than ever, for as traditional media like newspaper and broadcast television news continue to wither, a new form of media is flourishing across society, and the new form has engagement and opinion at its core.

11. Social Media

MARK
You know, you really don't need a forensic team to get to the bottom of this. If you guys were the inventors of Facebook you'd have invented Facebook.

-THE SOCIAL NETWORK (2010)

In the space of barely a decade, social media platforms have grown from a novelty to a global powerhouse. A service which in 2005 was mostly facilitating romantic interactions between college students in the northeastern United States was by 2011 a major player in toppling a repressive North African regime. Social media has changed politics, journalism, education, entertainment, and a host of other institutions, in ways both seen and unseen. The landscape of social media comprises a diverse collection of technologies, and that landscape is constantly evolving and shifting. Social media is, and will likely remain, both a changing force and a force for change.

The study of social media is a fertile and underexplored avenue of scholarship that is still in its infancy. The nature of the technologies that make social media possible also make it uniquely open to academic inquiry; at present scholars have only begun to scratch the surface. Social media allows the flow of information across vast networks of connected individuals, and how and why information propagates through those networks varies. A full theoretical treatment of the structure and function of social media networks is far beyond our scope in these pages,[1] though the interested student is encouraged to explore the basic tenets of graph theory, the branch of mathematics most relevant to social media networks.

Instead, our goal in this chapter is to identify a short list of the most important aspects of social media that the successful user must embrace. Though the language of social media is still rapidly evolving, we can clearly identify certain behaviors that most successful users share. Social media tools emerge, are embraced, and then abandoned at a rate far greater than traditional media tools, and different networks attract and appeal to different audiences. Rather than discuss any one particular platform, we attempt to identify those attributes which are shared across nearly all social media platforms.

It is important to note at the outset that social media is not the ideal tool for all working scientists. While these platforms can be tremendously powerful, by and large they do not lend themselves well to casual or occasional use. Modern social media users have come to expect certain standards of behavior as regards timeliness, responsiveness, level of engagement, etc. Together, these factors mean that success on social media requires a very significant investment of time and energy. For some, that time is better spent elsewhere. If you are someone who prefers analog to digital tools, if you keep your calendar on paper, if you check your email only once a day, if you prefer not to carry a phone, chances are that social media will not be for you. This is not a shortcoming, and should not be a source of shame; the existence of a technology is not a mandate for its use. Social media is most profitably exploited by those who are willing to embrace it fully, and integrate it into their lives to a degree which may make some users uncomfortable. By and large, it is not worth exploring social media unless or until you are willing to

---

1 The interested reader should consult *Networks, Crowds and Markets: Reasoning About a Highly Connected World* by David Easley and Jon Kleinberg. This engaging and comprehensive reference is frequently used as a text in graduate-level courses exploring social networks.

commit significant resources to making it a success.

Timeliness

The biggest factor that distinguishes social media from more traditional forms is also the attribute that makes it so powerful: social media happens in real time. The traditional newspaper model of overnight publishing seems a snail's pace compared to the lightning rapidity with which ideas propagate across the social media landscape. Indeed, most professional journalists now rely on social media platforms to provide them with the most current and up-to-date information, or to alert them to breaking news. This is because social media allows a direct and immediate connection between users, and the distributed nature of the networks allows for very rapid transmission. Events can be publicized and commented on literally within minutes of taking place.

As a user, this means that in order to be relevant, information must be timely. The tide of thoughts and ideas washes back and forth across social media on an hourly basis, and enthusiasm for trending topics dissipates as quickly as it accumulates. What's more, information on most social media systems is impermanent, as thoughts and ideas are quickly buried beneath new content. As a consequence, successful users must be both observant and responsive.

Successful observation of social media means frequent observation. It is not sufficient to check in to most social media sites on a weekly or even daily basis. Instead, it is useful to think of social media as an ongoing data stream, akin to a river. If you want to observe the objects floating by, you need to sit at the river's edge, and glance over at it regularly. Many successful users of social media have one or several of these streams or feeds open somewhere on their workspace throughout working hours, and modern mobile technology allows continued connection regardless of physical location. While it is not necessary to spend every minute of every day "plugged in", it is important to remain connected on a regular basis, so that when something of interest occurs, you are able to quickly respond.

Social media is a fast-paced and jumbled landscape, and users expect rapid responses more than polished ones. The traditional model of spending a week crafting a response to an article of interest does not work in the social media space. When something happens that you wish to comment on, or that is a relevant opportunity to reference yourself, your team, or your work, the most important thing is to respond quickly. The game is played in minutes and hours, not in days and weeks. While this rate varies somewhat based on the platform (a video sharing site or an online encyclopedia are less fast-paced than a news sharing network), the general rule is that faster is better. Social media is all about what is happening now, and you must be part of the conversation now if you wish to be relevant.

Engagement

The second great difference between social and traditional media is that social media offers far more interaction between producers and consumers of content. Indeed, one way to define a social media network is any technological platform that enables the sharing of so-called user-generated content. Unlike the traditional model of one source of information connected to many nodes, social media systems comprise many smaller sub-networks, with nodes constantly talking back and forth to one another.

For the user interested in connecting with an audience, this means that in order to be effective on social media, you must engage with that audience. It is not enough to simply post content and walk away. Social media encourages discussion and dialogue, and its users expect a high level of engagement. The lines between authors and readers are blurred if not erased, and a successful producer of content must engage and interact with her would-be consumers.

Note that this is distinctly different from the traditional media model, where the producer's job is more or less done once the content has been made available to the world. This approach will not work in the modern social media landscape. You must listen to what your readers, listeners, or viewers are saying, and engage them in conversation. While this may seem a daunting task, it is made much easier by the fact that social media allows you a much more accurate picture of who your readers, listeners, and viewers are.

Narrowcasting

The traditional mass media model relies on what we still call broadcasting: the transmission of a single signal to large number of receivers, with no knowledge or control over who is listening. While still a tremendously powerful system, its great shortcoming is that it is impossible to tell with any precision the size or composition of your audience.

Social media works in a fundamentally different way. Because social media systems are inherently peer-to-peer, the system itself keeps detailed logs of every interaction, and allows you to know not only how many people are accessing your content, but who those people are. Social media platforms vary in the level of access to this data they provide to the average user, but most allow a considerable amount. Even the most rudimentary analytics packages allow you to see the number of users, their geographic distribution, the technological devices they are using, and often detailed demographic information. You can see who visits your pages, how long they stay, and where they go when they leave.

This wealth of data is critical, because it allows you to carefully track your audience in a way that is impossible with traditional media, and ultimately to tailor your content to a specific audience. It may not be the case that you want or need a large audience, you often simply want to reach the right audience. As we said in Chapter Three, the correct audience for your message is the audience that can achieve your objective. Social media offers the tools to find and connect with that specific audience better than any other platform. If you write a weekly column that is only read by 100 people, but every one of them is the president of a university, it's likely a far more influential column than one read by 10,000 thirteen year-old boys.[2]

This process of identifying and targeting a very specific audience is what's called narrowcasting, transmitting a signal to a small number of very specific users. Social media is a great enabler of this technique, and it is one of the reasons that while traditional science journalism is dying out, science-based social media is flourishing.

The truth is that not everyone in society cares about science, or has the time and energy to pay attention to new scientific findings. Yet through social media, authors are able to connect to that segment of society that does care. We see in social media a deeper engagement with a narrower band of the population, and that is not necessarily a bad thing. What matters is that we are able to identify and connect with audiences that can achieve our objectives. Social media offers an unparalleled set of tools for defining those audiences, and a useful palette of metrics for measuring our success in connecting with them.

Crowdsourcing

Finally, social media platforms offer the capability for a widely distributed means of production known as crowdsourcing. Crowdsourcing is the generation of some resource (usually either money or information) in very small amounts from a very large number of people. This system is

2 13-year-old boys are both the producers and consumers of the majority of content on some social media systems, but they are not likely to be a major force in, say, decisions about international monetary policy.

not new, (the Golden Gate Bridge in San Francisco was famously funded through a series of small investments by the city's residents) but it is finding new applications as a result of social media platforms. Perhaps the best and most successful examples of this are online encyclopedias.

Though often derided in professional academic circles, online encyclopedias like Wikipedia are tremendously powerful and valuable tools. Open to editing by anyone, they have blossomed into some of the world's most used reference works, and despite the criticism leveled at them, the information they contain is of remarkably high quality. These massive, freely accessible, completely open repositories are becoming more and more frequently used in science, and the day may come when they begin to replace the traditional model for the dissemination of findings.

Crowdsourcing has also become an important tool for fundraising, allowing artists, tinkerers, politicians, and entrepreneurs to raise funds using small donations from a large population. Those fundraising efforts would not be possible were it not for social media. As the competition for public and private funds for scientific research grow more fierce, a small number of people in the scientific community are beginning to look at crowdsource funding as a possible alternative.

In closing, it is important to note that perhaps the most important thing to remember when wading into the landscape of social media is that ultimately, social media is just a technology. The people using it are still people, and as such, they will respond best to a compelling story. Find a way to frame your message as a narrative, and make the narrative structure clear to the audience. The technology may be different, but the underlying principle remains the same.

The American filmmaker Robert Zemeckis was once asked how new digital cinema tools would impact his craft. "We're still just telling stories around the campfire," he said. "All that's changing is the campfire."

12. Face to Face

```
                    LEON
The closer you get to being a pro, the
closer you can get to the client. The
knife, for example, is the last thing you
learn.

                    -THE PROFESSIONAL (1994)
```

We have saved our discussion of face-to-face personal communication for the final chapter of this section with good reason, for though it is one of the most common pathways of communication, it is the most difficult to master. In the same way that a true mastery of any art requires a rigorous and exhaustive study of the established forms as preparation for a complete abandonment of those forms,[1] face-to-face communication is most profitably explored only after one has become proficient in the more structured mediums discussed in the preceding chapters. Face-to-face communication is difficult for the same reasons that it's powerful: it operates through multiple channels, and it requires improvisation.

Face-to-face communication consists largely of speech, and as we discussed in Chapter Six, speech is powerful because it allows communication through several pathways beyond just the verbal information being conveyed. In face-to-face, one-on-one situations, all of those pathways are operating,

1 Examples include the study of formal landscape or portrait painting as a prerequisite for experimentation in expressionism, or training in classical music theory as a precursor to the exploration of jazz.

and more. Because you are engaged in a dialogue with just one other person, the importance of eye contact, body posture, and other nonverbal cues is heightened. The way another person smells, or shakes hands, affects your perception of them, and thus affects the way their message is received. A great example of this can be seen when one is speaking to another person through a translator. You never address the translator, you keep your focus on the person to whom you are speaking, and then wait to hear the translation. This is because the words are only one of the many channels through which you are sending and receiving information. Face-to-face interaction offers more simultaneously operating pathways of communication than any other mode of human communication, and the architecture through which it operates is so deeply embedded in human consciousness that messages are sent and received without either the sender's or the recipient's knowledge.

The situation is made inherently more complicated by the fact that face-to-face interactions are inherently unpredictable; you must operate without any foreknowledge of what you are going to say, or who you are going to say it to. It is dynamic, stochastic, even chaotic, and requires creativity, flexibility, and improvisation. It cannot be calculated in advance. Face-to-face communication is absolutely not a science. It's an art, and to master it, you must become a true artist.[2]

Our goal in this chapter is to give you a simple set of tools that will help guide you as you approach these interactions, but no textbook can make you proficient in this skill. Like any art, mastery comes only from years of practice. The principles outlined here are intended to help guide that practice.

---

2 More and more science communication training programs are using the techniques of improvisational theater, an approach pioneered by the Alan Alda Center for Communicating Science at Stony Brook University.

Start Talking

Perhpas one of the most difficult things for people beginning the practice of face-to-face interactions is learning to initiate and engage in conversation. This is especially true for students or junior colleagues, who may feel that they have little to offer in comparison to more senior people. This fear is almost always entirely unfounded. People in positions of authority are often anxious to hear thoughts and opinions that may be different than their own, as they can offer new insights and fresh perspectives. Despite the potential social awkwardness or any sense of discomfort you may feel, face-to-face contact is critical to success in almost any field, and you will never become proficient at it if you do not learn to start talking.

Ask Questions

When engaging in conversation, especially with someone you are not acquainted with, it is imperative that you try to learn something about your listener, and the best way to do that is to ask questions. Asking questions accomplishes two things at the same time. First, it shifts the focus away from you and toward the listener, which is much more likely to make them feel emotionally invested in the conversation. More importantly, it offers you the unique opportunity to understand who you are speaking to, and to shape your objective and thus your message to suit that particular audience.

In face-to-face conversation we have an audience of one, and that allows a unique opportunity to understand our audience in a way that would be impossible in any other medium. You needn't make assumptions about your listener's background, or their interests, because you can simply *ask*. Even the most basic inquiries, like: "What do you know about honeybees?" or "How familiar are you with photovoltaics?" will yield a wealth of information. If you are speaking to someone with no prior knowledge of your field, that will be a much different conversation than if you are speaking to a fellow researcher in your discipline. Especially when meeting a person for the first time, you need to find out what that person knows, and what they are interested in, and the best way to do that is to ask.

When asking questions, it is always better to ask open-ended questions, which require a more elaborate response than a simple yes or no. Consider the examples above. If the question were phrased: “Do you know anything about honeybees?”, and your listener simply says yes, you don’t have much to go on. By phrasing the question in an open-ended manner - “What do you know about honeybees?” - you are likely to get a more detailed picture of how much they do or don’t know. Crafting questions in an open-ended manner takes practice, but if you pay attention to it, it will eventually become second nature.

Lastly, don’t ask questions that you already know the answer to, and try not to ask questions that only have one “correct” answer. The risk here is that the person you’re speaking to will feel as though he or she is under examination, which makes most people uncomfortable. Your goal is not to quiz your listener, and rhetorical questions can make you seem haughty or aloof. Thus questions like: “Do you know what the most important factor in gene expression regulation is?” or “How big do you think the smallest artificial satellite is?” are usually bad choices. You should genuinely want to know something about your listener, so ask questions, and listen carefully to the answers.

Stop Talking

One of the most important thing to remember when speaking face-to-face with another person is that you must stop talking. We in the sciences frequently find ourselves in conversations about our work, and because our knowledge of that work is both deep and comprehensive, we could (and too often do) speak volumes about it. In order to be effective, conversation must be a two-way process, a dialogue in which both parties are speaker and listener. This can only happen if you remember to stop talking, and allow the other person to decide which way the conversation should go next.

This is particularly important if the conversation is highly time-constrained, as in the case of the so-called "elevator talk", where one only has the duration of a single elevator ride to make conversation. The temptation is to try and cram as much information as possible into your listener's ears, but precisely the opposite is true. In such situations, you must keep the conversation crisp and brief, and the best thing to do is answer questions as succinctly as you can, and then stop talking.

The reason this is so critical is that every time you stop talking, the person you are talking to gets to make a decision, to guide the conversation in the direction that is most interesting to them. Every time that happens, the course of the conversation changes, and each of those course corrections guides the conversation closer to one of the places where the other person wants to go, which should always be your goal. As we discussed in Chapter Three, the correct order of operations is to establish an objective, identify an audience, and then craft a message. When speaking one-on-one, the objective is almost always to leave an impression about yourself, and although counterintuitive, the best way to do that is to let the other person steer the conversation, tailoring the message to what they most want to hear.

Some people balk at this advice, concerned that if they do not fully articulate what they wanted to get across, an opportunity will be lost. The temptation is particularly strong when speaking to someone who is more senior or in some position of power over you, like the head of division at your company, or the director of a funding agency of your government. People fear that if they stop talking, and the person does not ask a further question, they will have lost the chance to convey the importance of their work. The truth is, however, that if the person you are speaking with is not engaged enough in the conversation to continue it, they aren't intellectually or emotionally invested, and all of your talking would have fallen on deaf ears. The power of face-to-face communication is the ability to convey more than just the facts. Face-to-face communication is about the intangibles, the human connection, and in order to focus on that connection, you must stop talking.

Why Trumps What

Generally speaking, it is better to speak less about what it is that you do (or have done, or hope to do), and more about *why* that thing is useful or important or significant. This is not something that comes naturally to most scientists. Because we devote so much time and energy to generating findings, we are often most enthusiastic about sharing those findings, about discussing what we know, or what we have shown. Yet the truth is that the vast majority of listeners (including many fellow scientists) will be far less interested in your findings, and far more interested in why your findings are important.

This is especially important when beginning a conversation, or when you only have a few moments to speak. Do not begin by discussing what you've done, begin by stating why you did it. This works because it instantly establishes a narrative proposition, with you (or your group, or your methodology) as the hero. Narrative is fundamentally ingrained in the human consciousness, and the minute you establish a narrative structure, your listener is drawn in. Start with your objective, start by explaining the *impact* of your work, and you will find your listener is far more engaged.

It is important to lead with the "why" questions, even if your particular experiment or line of inquiry has so far been unsuccessful, or if you have disproven your own hypothesis. Recall from Chapter Four that whether or not the protagonist achieves the objective is immaterial; it matters only that the audience understands the objective toward which the protagonist is striving. Dead ends and false leads are part of the scientific process, and you need not shy away from them. If you tell your listener why you were trying something, they are far more likely to be interested, even if you were ultimately unsuccessful.

Emotion v. Information

Talking less about what you've done, and more about why you've done it is essential because it more effectively engages a listener's emotions. As we discussed in previous chapters, this makes many in the sciences uncomfortable. We are taught to view emotion as a corrupting influence, as a contaminant which can spoil research. Yet the simple fact is that human beings (including other scientists) are far more influenced by our emotional reactions to things than we are by our intellects. Ours is not a species ruled by logic, or not by logic alone. To be successful in communication, we must embrace more than just the facts. We must embrace the audience's emotions.

Note that this is not to suggest that the facts don't matter. Far from it. Accurate, comprehensive, factual information is and will always be the foundation of good science. Information is absolutely necessary. But it is not often sufficient. This is particularly true when speaking face to face with another human being. The power of face to face communication comes from its ability to convey more than just factual information, and in order to use it successfully you must use those other channels to transmit some emotional content to your listener.

This is particularly important in the twenty-first century, when all of us are so deeply embedded in the Information Age. In this century, the cost of information is falling asymptotically toward zero. Facts are cheap. Almost all of us carry devices on our persons at all times that instantly link us to a global network of information. Factual data has never been easier to obtain than it is today. We are all afloat in a sea of information.

In this context, it is even more important to convey some emotional sense of the importance of your work to your listener. It is often a waste of time to spout statistics or quote figures when speaking to someone, unless those figures are being used to drive home a particular emotional point. The listener can easily find the factual information herself, and likely find the full breadth of your published work. All of that data transfer can happen elsewhere, outside the context of the conversation. What cannot happen is the transfer of emotion, the interaction which will instill in the listeners the desire to seek that information in the first place.

In the Information Age, if you find yourself having to choose between leaving your listener having more information about you and your work, or *wanting* more information, always choose the want. The information can be had anywhere, but the *desire* for that information can only come from you, and that is the thing you must focus on transferring to your listener.

The best way to do that, as always, is to employ a narrative. Talk about the *why* of your research, establish an objective toward which you are striving. Paint yourself as a protagonist engaged in some struggle. Briefly outline the obstacles in your way. Mention the stakes. In face to face interactions you can, in only a few words, cleanly paint a narrative proposition with yourself as the hero, and this, more than anything else, will leave an emotional (and thus memorable) impression in the mind of your listener.

Manage Expectations

When engaged in face to face interactions, it is important to manage your own expectations about the outcome of those interactions. Particularly in time-constrained situations, where you may only have a few minutes to try and express your key points, it is unlikely that you will affect some profound change in your listener. This is particularly true when we speaking with someone in a position of power. We run the risk of luring ourselves into the mistaken belief that if we can only clearly outline our points, if we make our case forcefully enough, that person will give us the funding, or regulatory permissions, or human resources we need. This thinking should be avoided, because it is both bad for us and bad for our listener.

A face to face interaction, especially a brief one, will almost never determine a major decision on its own, or on the spot. Big decisions require careful thought, reflection, an analysis of the relevant data, a weighing of the various options, etc. Someone who hears about a new research effort, or a new product, or a new process may be intrigued, but is unlikely to make a final decision about it right away. Expecting to get a final answer from your listener is unrealistic. What face to face interactions can do is arouse interest, or pique curiosity, or instill confidence. If your listener leaves the conversation feeling intrigued, and wanting to know more, that is a huge success. If you and your listener make a positive connection, you can easily carry on the conversation at a later date through some other medium. The goal of the face to face interaction should be to establish that connection, in hopes of furthering that communication.

Trying or expecting to achieve some result within the context of the conversation makes it more likely that you will alienate or annoy your listener. People are generally uncomfortable when pressured to make a decision or give an answer. The last thing you want is for your listener to feel badgered. As soon as the person you are speaking with starts to feel pressured or attacked, as soon as she is on the defensive, you've lost. A person who walks away from a conversation with negative emotions is very unlikely to seek to continue that conversation in another forum. Your goal should not be to get something out of your listener, but to give something to them. A conversation which ends with an exchange of contact information and a promise of further contact is a brilliant victory.

Practice

Lastly, as we noted at the outset, mastery of face to face interaction comes only through extensive practice. It is not something that will improve with study alone. Not everyone can become a master communicator, but everyone can improve their interpersonal communication skills, and the way to improve them is through practice. For scientists, by far the best way to practice is by speaking to people who don't know anything about science, and who don't necessarily care.

We in the sciences spend the vast majority of our time talking with fellow scientists, or with the people who fund, regulate, administer, or write about science. Thus we sometimes mistakenly believe that most people in the world know something about science and why it's important, because those are the people with whom we interact. Yet the truth is that whole industries and huge swaths of the population have little or no contact with or interest in science.

That segment of the population makes the ideal audience for honing your skills in face to face communication, because they are the perfect test case. The biggest challenge in face to face communications (and, as we have seen, in all communication) is tailoring your message to achieve the desired impact on your audience. When dealing with an audience that has zero foreknowledge of your field or why it might matter, you are operating at the limit of this challenge. Finding ways to make your work accessible, crafting narratives about your work with sympathetic objectives, or choosing language about your work that can be universally understood are all more difficult when your audience is hearing about your work for the very first time, and because of that difficulty, it is infinitely more rewarding.

Speak with students at a local school. Speak with members of community groups or churches. Speak with the stranger who sits next to you on the train or airplane. Speak with your neighbors, speak with your friends, speak with your own family. But speak, and speak often. You will likely struggle, will find yourself frequently misunderstood, will face strange questions or blank stares. You will be uncomfortable, and all of it will make you better. As in food, just because it tastes foul doesn't mean it's not healthy.

Through the process of speaking with people who know nothing of your work, you will slowly find the explanations that work. You will find the descriptions that enlighten. Most importantly, you will identify the narratives that resonate. By working in the limit of unfamiliarity, you will develop a sense of how well your listener understands. You will learn to stop talking, and ask questions, to concentrate on emotions, and to manage your own expectations.

Speaking with people from outside the sciences teaches discipline, enlarges skill, and hones instinct, all of which are necessary ingredients in the mastery of an art. You will never be perfect, but practice will make you far better. If you are proficient speaking with people from general publics, you will be infinitely more effective when speaking with people within science.

Part Three

Consequence

13. Controversy

```
                    HOWARD
I want you yell: "I'm as mad as hell, and
I'm not going to take this anymore!"

                           -NETWORK (1976)
```

Anyone who regularly communicates about science, and especially those who seek to engage public audiences about scientific topics, eventually must grapple with one of the most complex and thorniest issues in the field: the question of how to address scientific controversy. Controversies come in all sizes: they may be as small as a disagreement about the importance of one particular species in one particular region of the world, or they may be as large as the origin of the universe or the fate of this planet. Controversy, more than any other aspect of the scientific enterprise, arouses people's passions, and the question of how to communicate about controversial issues is neither easy, nor new.

Scientific controversy has been with us for as long as science itself. The first philosophers in Hellenistic culture were not universally agreed with (certainly not by the common citizen), and yielded far less influence in their own day than the poets or priests who were their contemporaries. Throughout the Middle Ages, science (as least in the west), languished for nearly a millennium. The late Renaissance is rife with tales of science suppressed or ignored, and from the Enlightenment to the dawn of modernity, there was a constant struggle between the new sciences and the superstitions of old. Even in modern times, scientific controversy remains a fact of everyday life, in part because controversy is a necessary part of the scientific process.

Often, however, when scientific controversies spill over into public discourse, or when scientific questions or issues enter the realm of public policy, those of us trained in the sciences strain against what we view as public apathy or public ignorance. Too often we see arguments on the other side of a debate that employ pseudoscience, or conjecture, or clever rhetorical tricks, or outright damned dirty lies. We scientists shake our fists in fury and pound our heads, secure in the knowledge that the facts are on our side, but nevertheless losing the argument. We see this problem again and again, in many different places, and on many different scales.

In this chapter, we hope to present a lens through which to view the origins of this problem, and propose a method to address it. First, we discuss public perception of science, and what science is and is not. Second, we return to our discussion of narrative structure, and discuss in more detail the concept of narrative reference frames, and the role they play in people's perceptions.

What Science Is

The simple question: "What is Science?" is probably one of the most important (and least examined) inquiries that affect the way scientific findings are interpreted and understood by broad audiences. Even those of us who spend our entire careers as working scientists infrequently take the time to really investigate this question in depth, preferring to leave the matter to philosophers. Yet it bears deeper examination, for if you wish to engage in public discourse about science, you must first have a firm notion of what science really is, so that you can begin to understand how the general public might perceive it. Though science may be many things, its first and most obvious aspect is also the one with which general audiences are most familiar: science as a body of knowledge.

Science As a Body of Knowledge

Acquired over the course of the last two thousand years, the depth and breadth of factual information that science has accumulated is as dazzling at it is diverse. We discovered and described species from the densest jungles and the deepest oceans. We have tracked the motions of the stars, the planets, and the distant galaxies. We have plumbed the smallest scales of matter, and learned what makes materials behave as they do. Only recently, we are beginning to gain a true understanding of the inner functions of biological systems, including the working of the human body and the human mind. We have discovered the remains of creatures more awesome and terrifying than anything from ancient mythology, and we have walked on the surface of the moon. From fire to fiber optics, and buoyancy to the Big Bang, science now knows many, many, many things.[1]

It is this aspect, our accumulated scientific knowledge, that most people most closely associate with the scientific enterprise. This association may stem from the way in which science is most frequently taught; where students are asked first to learn and then remember a laundry list of things that other scientists have learned throughout history, be they the names of the stars and planets, or the periodic table of the elements, or the component parts of the cell. When most people think about "learning science," they think of acquiring some portion of that knowledge.

Yet this aspect of the enterprise, our collective scientific knowledge,

---

1 Incidentally, it may be the success of this enterprise that creates some of the difficulty we now face in sharing science with others; because there are so many things a good scientist must know, she has little time to study anything else. Every semester spent immersed in the equations of Schroedinger is one where the poetry of Shakespeare and the philosophy of Schopenhauer are left aside.

is only part of what science is. A student could spend years acquiring a substantial portion of this knowledge (and many do), without ever really learning to think and act scientifically. This is because science is not just a body of knowledge, not merely a catalog of things that are known. Science is also a way of knowing things, a methodology for learning about the structure and operation of the things we perceive in the world.

Science As a System of Thought

The second (and by far more important) aspect of science is the way it teaches us to investigate the world.[2] Often introduced to students as the "scientific method," this system prescribes a rigorous approach to observation and experimentation, and emphasizes the development of testable models. It is the method, far more than the body of knowledge, that has given science the immense power that it possesses today, for throughout science's history, we have stumbled upon new behaviors or new phenomena which did not fit into the existing catalog of established knowledge, and through the method of investigation, found new and different ways to describe and explain those phenomena.

It is this second meaning, science as a system of thought, science as a process rather than a product, that those of us in the sciences most often mean when we use the word "science." When we think of ourselves as "doing science," we think of designing and conducting an observation or experiment. We recognize that the body of knowledge generated by science is not static, that new information constantly reshapes our understanding, and that even the most long-held beliefs in science are subject to change as a result of new findings. We scientists embrace the idea that science is fundamentally anti-authoritarian, that it rejects all forms of dogma, that our

2 An excellent popular treatment of the scientific process and how it contrasts with superstition and pseudoscience is *The Demon-Haunted World* by Carl Sagan.

current knowledge must constantly be challenged. We question everything.

The great irony when it comes to scientific controversy is that we scientists, who make it our business to constantly challenge the findings of those in authority, too often ask various publics to accept the validity of our own findings, based on our own scientific authority. We ask our fellow citizens to accept science unscientifically. Too often, far too many of us, rather than do the hard work of making the evidence plain enough for anyone to see and understand, simply state our certainty about what that evidence proves.

This approach has hypocrisy at its core, and though it may seem expedient, in the end it is injurious to ourselves and to society. For if we ask those around us to accept the truth of what we say based on the faith they place is us, we are asking them to abdicate their own ability to think scientifically. We are eroding the critical thinking capability of society at large. Consider the citizen who is asked to believe that science is true, because the learned men in the university have written many books that say so. How is that different from her being asked to believe that certain religious beliefs are true, because the learned men in the seminary have written many books that say so? If we wish to foster a society in which science is truly embraced, we must encourage the embrace of both science as a body of knowledge and science as as system of thought. We must ask our fellow citizens to question everything, including the things we tell them. We must persuade them with evidence, not with argument.

Narrative Reference Frames

The other great difficulty that arises when dealing with scientific controversy is one of narrative reference frames. As we've discussed, narrative is an integral part of the way people view the world. People subscribe to narratives that arc across their entire lives, and interpret new information differently based on their own narrative reference frame. In the same way that physical events vary based on reference frame (the apparent size or speed of an object varies with the distance and speed of the observer) so too do factual events appear differently based on the frame through which they are viewed.

Perhaps the clearest illustration of this effect can be seen in the world of sport. Imagine a sports stadium in which there are two teams, and two sets of fans. The seating is segregated so that all the fans of one team are on one side of the stadium, and all the fans of the second team on the opposite side. The first team is trying to move the ball into the goal, and the second team stands in their way.

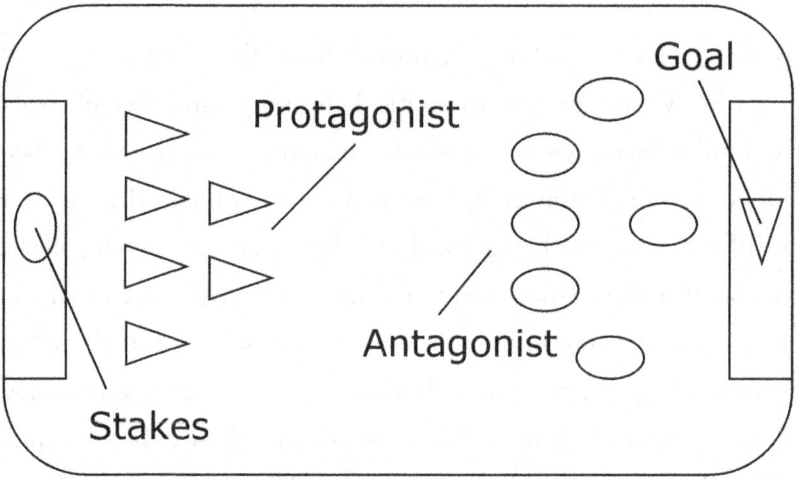

Figure 13-1: Sport as story.

The narrative proposition here is immediately apparent. For fans on the near side, their team is the hero, striving toward their objective (the goal), and the other team is the obstacle that stands in their way. Someone wants something badly and is having trouble getting it. The stakes are that the other team might score goals, which would, from one side of the stadium, constitute a tragedy.

Now consider the same game, viewed from the other side of the stadium. From that perspective, the narrative is precisely the opposite. Those fans view their team as the hero, striving toward their own goal, and the opposing team as the obstacle. Because of the shift in narrative reference frame, the roles are inverted, one side's protagonist is the other side's antagonist, and vice versa.

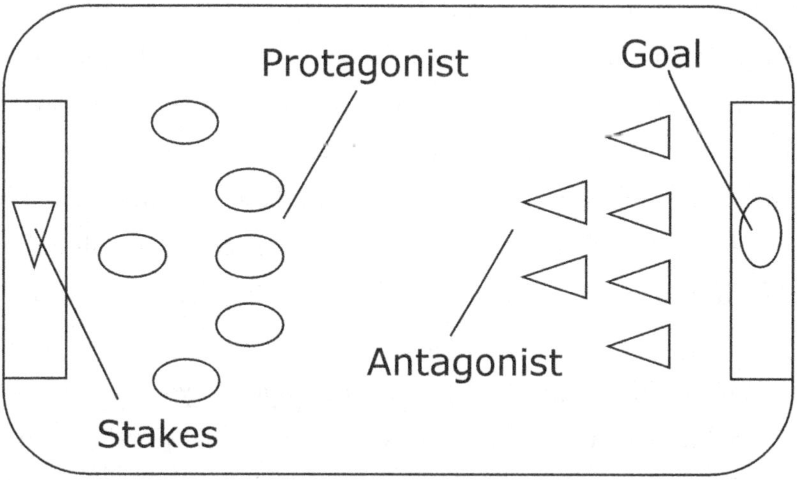

Figure 13-2: The same game viewed from the other side of the stadium. The facts are unchanged, but the narrative is reversed.

The important thing to note here is that all the fans are watching the same game. The facts are never in dispute. The two groups of people are viewing the same events at the same time, and they would not disagree with one another at all about what has transpired, but what seems to one group like a brilliant victory and a cause to rejoice will seem to the other like a heartbreaking defeat. Because the two groups entered the stadium subscribing to different narratives, their emotional reactions to the events on the field will be diametrically opposed to one another.

It is this divergence of views as a result of narrative reference frames that so many of us fail to appreciate when dealing with scientific controversy. The larger narratives about science, and the role it plays in society, are not transient ideas. They are deeply entrenched, and people acquire them over the course of their lives. The narrative reference frame to which a person subscribes is a function of many factors, and may be influenced by their culture, their religion, their level of education, etc. These are not positions that are taken lightly or that will be quickly or easily cast aside.

The trouble grows infinitely more complicated when speaking to a group of people who have come to believe that science itself, or those who practice it, are in some way the villain in a larger narrative. If you have been raised to believe that the goal of your life should be to achieve spiritual purity, and that one of the obstacles that stands in your way are the lies told by the men and women of science, that narrative will shape your understanding of factual information. When a scientist offers more facts, you will only hear more lies, because that is your context for processing that information.

This is the most important concept that you must keep in mind when dealing with a controversial scientific topic. If you have chosen to engage a public who may be hostile or suspicious of your position, you must identify that public's predominant narrative. Understanding the narrative proposition that shapes their understanding will help guide you in crafting a counter-narrative and identifying an impact that is achievable. It is not likely that you will convince a group of people to abandon the kind of large narrative that shapes their lives.

Instead, you must try and focus on offering some compelling alternative narrative. We in the sciences too often get wrapped up in the belief that we need to change people's minds, to convince them of the truth of what we say, to win them over with science. These efforts, though well-intentioned, rarely succeed. Instead, we should concentrate on what it is that we want people to do. It is often the case that there may be another more compelling reason for people to behave in a certain way, and if we look for a narrative that reinforces that reason, we will be more successful.

When engaging in scientific controversy, especially when dealing with an entrenched opposition, don't worry about changing what people believe. Focus instead on how you think those people should act, and you are much more likely to be successful. Examples of this principle in action, and a further discussion of competing narrative reference frames can be found in Chapter Sixteen.

14. Public Engagement

DOROTHY
Toto, I've a feeling we're not in Kansas anymore.

-THE WIZARD OF OZ (1939)

Once upon a time, excellence in research was the only requirement for excellence in science. Researchers in universities, or in laboratories sponsored by government or industry, were free to investigate the questions that interested them most. Funding was plentiful, and easily obtained; scientists were free of the fetters of oversight or the meddlesome interference of regulations. All that was required was that we find interesting questions and produce compelling answers, which would be duly reported in the literature for our fellow scientists to read. We were never required to wade out among the great unwashed masses and explain what we were doing, and those few of us who chose to do so were treated as pariahs; as men and women who had forsaken their sacred vows to the monastery of the scientific enterprise.

That, at least, is the legend that many scientists of the previous century recount. Whether that halcyon past ever existed is dubious, and even if it did, it is doubtful that the science it generated was any better than the science done a century before, or in this century. Regardless, the legend is markedly different from the reality of science today.

More and more, scientists are coming to realize (and those who fund science have begun to mandate) that the value of scientific research cannot be measured solely by its intellectual merit. Researchers today are increasingly required to demonstrate that their work has some tangible value to the rest of society, and though that value may come in many different forms, it very often involves some sort of public engagement.

Public engagement with science is a relatively new concept, and it has replaced what social scientists previously referred to as "public understanding of science." The previous model focused on a one-way flow of information; on the the idea that science had a responsibility to inform various publics, that there was some deficit that needed to be corrected. Public engagement is a broader concept, and embraces the idea that the relationship between science and society is more complex, that while it's true that people should know more about science, it's also true that science should know more about people. The theory behind public engagement is that a true dialogue between science and society improves them both.

In this chapter, we present some of the formats through which public engagement with science can profitably take place, and briefly discuss some of the most common objectives that public engagement can achieve. This is by no means an exhaustive list, but is meant to serve as a representative sample of the kinds of activities available to the working scientist, and the kinds of impacts that one can hope to have on various audiences.

Informal Science Education

One can engage various publics in a host of possible formats, and through a broad collection of possible mediums, ranging from science cafes to mobile games, and from museum programs to community clubs. Nearly all of these interactions fall into the general category of informal science education, broadly defined as any engagement in science that happens outside of a formal learning environment like a classroom. Research has shown that the vast majority of interactions most people have with science happen in these informal environments, and that the quality of those interactions plays an important role in how people view and understand science and the scientific process. Informal science education is a broad and diverse enterprise, but the most important institutions in that field are museums and science centers.

Informal science education as a field emerged in the 1960s, when a new generation of institutions began to challenge the traditional model under which museums had operated.[1] These institutions moved away from the concept of museums as warehouses for the collection and storage of artifacts, and began to embrace the notion of interactive experience, of building exhibits which enabled visitors to engage with them, to manipulate them, and through that manipulation (it was hoped) to learn something about the effect on display. This was a marked departure from standard practice in museums, where objects were displayed in glass cases, and "Do Not Touch" was the guiding mantra throughout. The new centers were just the opposite, places where audiences were visitors were not just allowed, but encouraged, to touch things, to explore, to play.

This was a new approach to presenting science, and it was successful partly because it allowed the visitors to play the role of scientist, to explore and discover phenomena on their own. The new model quickly spread, and today it is by far the dominant model throughout the world. Even most traditional museums of natural history have re-invented themselves along the lines of the new model, and though they still have huge collections of artifacts, they focus more on creating and delivering these interactive experiences to their visitors.

Far beyond just the exhibits and programs that happen inside the

---

1 One of the most important of these was the Exploratorium, in San Francisco. The story of that institution, and its founder, Frank Oppenheimer, is beautifully told in *Something Incredibly Wonderful Happens*, by K.C. Cole.

building, museums and science centers play an increasingly important role in engaging public audiences with science outside of the museum, and in furthering the scholarship of science communication and science engagement. The largest institutions, like universities, are true research centers, where the question of how best to engage publics with science is being studied and explored.

For the working scientist, these institutions can be an invaluable resource. Though their original goal was to educate visitors about basic scientific concepts, most of which were already well-established by the end of the nineteenth century, more and more these institutions have come to embrace current scientific research, and have begun to serve as important platforms for publics to engage with scientists around these issues. Whether it is through public forums, on online discussions, or science cafes, engaging public audiences is central to the mission of museums and science centers. Many offer resources for scientists about how to be part of that engagement, and can serve as an important liaison between researchers and their relevant publics. There is a long list of activities that informal science education institutions offer. For the purposes of illustration, we will present three of the most common: public lectures, tabletop demonstrations, and science cafes.

Public Lectures

One of the oldest and still most effective forms of public engagement with science is the public lecture. Designed to present scientific concepts in an accessible, engaging way to audiences of all ages, the public science lecture has a long and storied history. Its earliest ancestor is likely the Christmas Lectures at the Royal Institution in London, which were begun by Michael Faraday in 1825, and continue to this day.[2] The Christmas Lectures began

2 The Royal Institution makes lectures from past years available on its website;

as a lecture series for young people, at a time when there was little formal education available, and the science being presented was still in its infancy. Over their long history, that lecture series has evolved into a slickly produced television event, now seen around the world, but the fundamental premise is unchanged; they consist of a single scientist presenting scientific concepts in the most general way to a room full of (mostly young) people.

There are now public lectures about science held in institutions around the world, and at all times of the year, and most of them are built on the model of the Christmas Lectures. One of the reasons for their continued popularity is that audiences seem never to tire of hearing about science from actual scientists. The best public science lectures are not given by actors or professional public speakers, they are given by working scientists, because there is something about hearing from a real live scientist that audiences find compelling. Public lectures can be a powerful format for engaging with a public audience, but in order to be successful, they require not just good science, but good storytelling.

The real challenge when preparing to give a public lecture is the great uncertainty about the audience. Whether the lecture will take place in a museum or science center, in a university lecture hall, at a town hall, or as part of some larger festival or event, it is almost impossible to predict in advance who will show up to fill the seats. Chances are high that the audience will be very diverse, comprising a wide range of interest in and understanding of science. It is thus essential that the lecture employ the language that all audiences intuitively understand: the language of narrative.

Whether the topic is geology or physiology, whether you are talking about space travel or neurosurgery, it is essential that you craft a clear narrative proposition. Because these lectures frequently center around science that is already very well established, one of the most common techniques is to draw

---

they can be an excellent source of inspiration for someone preparing to give a public lecture.

a narrative from history. The science under discussion didn't develop in a vacuum, it emerged over time as the answer to a question or the solution to a problem. The question or the problem was trouble to someone, and the solution was something they wanted badly. Find that narrative and use it.

The other hallmark of the public lecture is the demonstration, they place where the lecturer leaves off simply *telling* us about the science, and *shows* us something instead. This tradition extends back even further, with roots in the early scientific societies of the 17th century (discussed in Chapter 7). The design of a good scientific demonstration is an art in its own right, and a complete discussion is beyond the scope of this book.[3] For our discussion here, it suffices to say that every good lecture should have at least one (and preferably more). Many of the best have been around for decades, so don't be shy about finding a good demonstration documented elsewhere and using it as your own. Even though it is familiar to you, it will likely be new to the audience, and that will make it effective.

The last thing to remember about preparing a public lecture is to start early, solicit feedback, and heed whatever advice you receive. Many scientists spend much of their careers either speaking to fellow scientists, or lecturing to students of science, both of whom have a vested interest in the content. With publics, this is not the case. The lecture must hold the audience's attention on its own merits. Especially if you are venturing into this territory for the first time, it is therefore imperative that you ask for input about what will and will not resonate with your likely audience. Professionals who work in informal science education can be invaluable to this process. Their expertise gives them a perspective that many scientists lack, and they will be able to tell you what will work with an audience, and what won't. If you have access to one of these individuals, ask their advice, and take it.

---

3 Large institutions often have a full-time professional whose only job is the design and maintenance of apparatus for science demonstrations.

Tabletop Demonstrations

One of the other mainstays of informal science education is what is known as the tabletop demonstration. Also referred to as "cart demonstrations," these are small-scale demonstrations of some scientific phenomenon that are conducted in a face to face setting between one facilitator and one or more participants. The tabletop demonstration is much less structured than a public lecture, because it allows for a dialogue between the facilitator and the participants. The general idea is that the phenomenon on display will serve as an entry point for the participant, encouraging her to ask questions. The strength of the tabletop demonstration is its portability and flexibility. Because they can be performed in almost any setting, they are frequently employed in travelling programs, at fairs and festivals, during after school programs, etc.

Tabletop demonstrations are frequently structured with a simple script. First, the facilitator demonstrates some behavior or phenomenon to the participants. Next, the facilitator suggests an alteration to the experiment they just saw, and asks the participants to predict how the behavior will change. The facilitator then repeats the demonstration with the alteration. (Good tabletop demonstrations usually display an unexpected or unusual behavior). These interactions are powerful for the visitors who experience them, because they are recapitulating the basic act of scientific investigation. They observe behavior, alter a variable, predict results, observe behavior again, and then try and make sense of what they have seen. This experience has the ability to produce a powerful sense of wonder in the participants, and often leads to further questions, which is the ultimate point of the exercise. This system works because it creates a narrative in which the participant becomes the protagonist.

Though many of the most common tabletop demonstrations focus on basic principles from physics, chemistry, and biology, it is possible to construct demonstrations that serve as an entry point to discuss current scientific research. Though much of science today is conducted at time or size scales that make direct observation difficult, it is usually possible to find an observable behavior which at least analogous to the behavior under investigation in your lab. Finding and crafting a tabletop demonstration based on your research may not always be easy, but is almost always possible.

Science Cafés

A forum for public engagement whose popularity has exploded in recent years, the science café (also known as a café scientifique), is a gathering that takes place in a casual setting like a pub or café, in which one or more scientists talk about a topic in science. These events are unique in that they tend to attract almost exclusively adult audiences, and they tend to focus on topics related to current issues in science and technology, rather than on basic scientific principles.

The format varies based on location and organizer, but the typical model involves a brief presentation by one or more scientists (usually no more than 10-15 minutes), followed by a discussion period, in which audience members ask questions or give feedback based on what they have heard. The discussion period is usually moderated by a host (often an educator or journalist) who also introduces and concludes the event.

This simple notion, the idea of scientists talking with people in places where people normally congregate, has proven incredibly powerful. A relatively recent invention, the science café has quickly become one of the most popular formats for public engagement with science. People who are interested in scientific topics, but who might feel intimidated by the formality of a lecture in an academic context, find themselves able to ask the kind of questions that matter to them most. Experience has shown that there is a significant public appetite for these events, and that people want to hear what scientists have to say, but don't feel like they have access to them. The science café, because it takes place in a casual and welcoming environment (and perhaps because it usually involves the universal language of food and drink), seems to lower people's inhibitions, and fosters an environment of honest, open dialogue that is seldom found in other contexts.

The organizers of science cafés are usually looking for scientists who are willing to participate, and they offer a tremendous opportunity both to engage with your local public, and to hone your communication skills. While these events can be intimidating, they can also be transformative, and success is more likely if you follow a few simple guidelines.

First, if preparing to speak at a café, talk with the organizer about the topic, and about the issues most likely to come up during discussion. Familiarize yourself with these issues, and be prepared to offer an opinion, even if the question is outside your particular specialty. People don't come to these events just to hear about your particular science, they come to hear your opinion as a scientist. In this forum it is appropriate to express your personal views, so you needn't constrain your comments to things that you have data to support.

Second, remember that you will be engaging people who may have views that are starkly different than your own. Just because they disagree does not mean they are unintelligent or poorly informed. Be respectful of others' opinions, and do not engage in back-and-forth argument. If you feel an audience member is being combative, let the moderator deal with it. If you get drawn into a debate with one audience member, you risk compromising your credibility with the rest of the audience.

Lastly, try to paint your remarks in the context of some larger narrative. As discussed in Chapter 5, the types of broad issues in science and technology that will most interest adult audiences have some narrative structure, often with science or society as the protagonist. Embrace one of these narrative frames, and use it as the backdrop for your remarks, but keep in mind that your audience may be viewing events from an entirely different frame, as discussed in Chapter 13. If you are unable to find a larger narrative context for the subject at hand, then tell a simpler story with yourself as the protagonist. Tell us what you are looking for in your work, and why you want to find it so badly, and you will instantly have an audience that is emotionally invested in what you have to say.

Impacts

Public engagement with science is not easy, and it's not always required. Many scientists go their entire careers without engaging any public in any meaningful way. Though some institutions and agencies have begun to mandate some form of public engagement, it remains tertiary to the careers of most working scientists, and the researcher who truly engages in a deep and meaningful way with some public remains the exception rather than the rule. Yet this is slowly beginning to change, and the latest generation of scientists seems more interested in public engagement than any that has come before.

This change is not just a benefit to society, but a benefit to science itself. Much of the literature about public engagement focuses on how that engagement improves society: by increasing science literacy, by ensuring that a sufficient number of young people enter the science pipeline to provide a scientifically trained workforce, and by fostering broad public support for science that will translate into better public policy. These things are likely true (at least to some extent, social scientists can and do argue about degree), but regardless of the benefits to society, public engagement is beneficial, because it makes for better scientists, and better science.

Public engagement with science is not just about science providing information to society, it's about society shaping the kinds of questions that science asks, and deciding what to do about the answers. The scientist who engages with publics is improving her communication skills, her artistry as a storyteller, and her ability to capture attention and stimulate imagination. But she is also learning what the people with whom she engages care about, what they hope for, what they fear. Engaging a public helps scientists see the world as those outside of science see it, and that added perspective ultimately helps shape the science itself.

Too many in the sciences fear public engagement, and avoid it at all costs. They are doing a disservice to themselves, to their fields, and to society at large. Public engagement does more than help people understand science. It helps science understand people. And science shaped by a broader and deeper understanding of the human condition is better for all.

15. Public Policy

MARULLUS
You blocks, you stones, you worse than
senseless things!
O you hard hearts, you cruel men of Rome,
Knew you not Pompey?

-JULIUS CAESAR (1599)

There is perhaps no arena of human activity more important to science than public policy. The legacy of the twentieth century is that almost every nation on earth now embraces and depends upon both scientifically-enabled technology, and democratically elected government. Neither the great discoveries of science nor the great feats of technology that humankind has accomplished over the last century would have been possible without the integral participation of governments. Today, government entities and agencies remain science's largest and most important benefactors. As society becomes more and more dependent on increasingly complex technologies, governments must make difficult choices about how to oversee the use of those technologies, and science should be one of the factors that guide the decision-making process. In the 21st century, science and democracy depend upon one another.

Despite the symbiosis between science and policy, practitioners of the two crafts are notorious for their inability to understand one another. Both are highly specialized, have long histories, and have traditions, customs, and languages of their own. Thus travellers from each sphere find themselves lost in the other, stymied by practices that seem archaic and structures that seem byzantine. Scientists are as uncomfortable negotiating policy as politicians are doing science, each finding the other's work maddeningly slow or hopelessly irrelevant.

The study of the intersection between science and technology and public policy is a vast academic discipline of its own.[1] There are several journals, countless books, and numerous schools and institutes devoted to it. Even a survey of that landscape is beyond our scope here. Rather our goal in this chapter is to highlight the one key element of policymaking that is most germane to the theme of this book: narrative.

Policy & Narrative

Public policy takes many forms. The development and enforcement of regulations, the establishment and oversight of agencies, the appropriation of funds, and much more all fall under the umbrella of public policy. Many of us who are technically trained (or simply scientifically minded) naturally assume that the underlying process of developing these policies closely mirrors the process by which we would approach a technical problem. We imagine that policies emerge as the result of identifying a problem, establishing the parameters by which we will measure whether or not it has been solved, making an exhaustive list of possible solutions, evaluating each solution's strengths and weaknesses, and then selecting the optimal solution as determined by the evaluation. In short, we think of policy-making like any other optimization problem; ultimately the answer is in the numbers.

In reality, however, this is rarely if ever the case, for public policy is decided by people, and people are never swayed by numbers alone. As outlined in our discussion in Chapter Two, the Apollonian mind, the rational,

1 The field is known as STS, for either Science, Technology, and Society, or Science and Technology Studies, depending on the source.

logical, quantitative part of our consciousness may be able to convince us that something is true, but it is the Dionysian mind, the instinctual, qualitative, emotional psyche that compels us to do something about it. It is therefore the case that most (if not all) public policy emerges from some stimulation of the Dionysian mind, from some emotional response, and those responses are most frequently triggered through the use of storytelling. One often hears talk of the so-called "policy narrative," how elements of public policy are portrayed, in order to sway public opinion, but the relationship is deeper than that. Narrative is more than just a structure used to communicate policy once it is crafted. In many cases the narrative process is what defines policy; much of public policy *is* a narrative.

To see how this might be so, consider an early pre-legal society that has not yet developed law, e.g. a farming community in Ancient Greece. (It is likely not an accident that the first people to codify storytelling were the same people who first codified democracy). One day it is discovered that some rogue has assaulted a farmer on his way to market, murdering him and stealing his harvest. The other farmers, seeing this, and concerned for their own safety, get together and decide that in their community a man who kills another man and takes his property shall be captured and put to death. In short, they pass a law making murder illegal. It is not difficult to see the narrative proposition here. The farmers are someone, and the something they want badly is to travel safely to market. The murderer represents potential trouble they may have, so they fashion their policy as a means to overcome him.

Simple though this example may seem, history tells us that law did in fact evolve this way. (Criminal codes are some of humanity's oldest laws). Societies do not sit down and conjure up full legal systems out of whole cloth; instead, they emerge piecemeal over time, with each new piece of legislation written to address a specific problem or redress a specific wrong. In each of those cases, there is some underlying narrative proposition, someone who wanted something badly and had trouble getting it. We find this throughout history and on all scales of government, from huge international agreements affecting billions to the most mundane ordnance on a local municipal council. Public policy most often emerges to help someone overcome some trouble and obtain something. Policy *is* narrative.

This is likely the source of much of the disconnect that scientists feel when dealing with questions of public policy. We want to be able to reduce policy questions to analytical questions, to solve problems with data. Yet we forget that the whole point of science is that is independent of any human bias. Scientific findings are only valid if they are independent of the observer; any other human should be able to perform the same observation and experiment and reach an identical conclusion. This is the polar opposite of how policy is constructed. Public policies are entirely dependent on human bias; they exist for the purpose of affecting groups of people. And while numerical analyses can and should be part of effective public policy, we must remember that those policies exist to serve the wants and needs of groups of people. Policy exists because someone wants something, and has trouble getting it.

Science & Policy

Embracing the notion that policy is fundamentally a narrative is essential when using scientific arguments to influence policy decisions. Science can and should be a voice that informs public policy, but it will never constitute a complete policy position. In order to be effective in policy discourse, science must provide an argument that supports or advances a narrative. To illustrate this, consider a hypothetical example:[2]

A small shorebird, which makes its nest on the beaches along

2 This scenario is not purely hypothetical, it is drawn from actual controversies surrounding *Charadrius Melodus*, the Piping Plover, and the protection of its habitat on beaches along the northeastern coast of the United States.

the coastline of a heavily populated region, has become an endangered species. Destruction of habitat and interference with its nesting sites have made the species' survival uncertain. Protection of the species requires extensive protection of those sites during certain times of the year, which requires restricting human access to the beaches, which are popular tourist destinations. Tensions between local residents and government regulators are rising, as some communities are chafing against the beach closures. To help ease the tension, a conservation biologist has been sent by the government to address a community meeting where the controversy will be discussed.

The typical biologist will arrive at the meeting well armed with information. She will discuss the bird's behavior, and its breeding. She will show charts documenting the decline of the species in recent years. She will have maps that indicate known nesting sites, and how they must be protected. She will have results from numerous field studies demonstrating the efficacy of current protection schemes, and she will have historical data that clearly show that unprotected beaches cause the destruction of habitat. She will argue, indeed, she may conclusively prove, that the closure of local beaches, inconvenient though it may be, is absolutely essential to the protection and ultimate survival of the species.

Yet even if she has done this, she will have failed to put forward a convincing policy argument. For at the end of her remarks, a local resident might reasonably ask: "Why don't we just kill all those birds?"

Our biologist would likely stammer at this, or begin furiously blinking. The question is one she is unprepared to answer, both because it is fundamentally not a scientific question, and because it is completely outside of her narrative frame of reference. And yet it is, ultimately, the policy question under discussion. Why not let the species go extinct?

There are, of course, a host of excellent, compelling answers to this question. Some of those answers are based in science (the preservation of food webs, etc.), but most of them are not. Most of them are based in fundamental ideas about the relationship between human beings and the natural world. These are questions of emotions, of ethics, of moral principle. Our biologist should have begun her remarks, indeed spent most of her time, answering these questions, for these are the question from which a narrative emerges. She must establish that we as a society (the protagonist) want to protect the birds (our objective). The science then becomes an argument in support of that objective. This context is critical, and it is the thing that scientists so frequently fail to provide.

Policy & Science

The careful attention paid to narrative is important not merely when science is guiding policy, but also when policy is guiding science. Scientific research in almost every field is supported to one degree or another by public funds, and the continuation of that support is by no means guaranteed. Support for science, like almost all public policy, has a narrative behind it.

In the case of applied research, this narrative is clear, and oft-repeated. The benefits of applied research are easy to see, and the dividends it pays to society, through better health, a stimulated economy, a better-protected environment, etc. are things that most people can easily agree we all want.

The case for basic research is a more difficult one to make, however, which is one reason that in some countries support for basic research seems to be on the wane.[3] Yet there is a narrative to be told. In the basic research narrative, the thing that we as a society want is knowledge itself, a finer-grained understanding of how phenomena in the universe work. We want this knowledge because we know that learning things about the world, even

3 "Basic" research would be easier to sell if we called it something else. For most people, the word "basic" connotes something rudimentary, crude, or unsophisticated. "Fundamental" research would be a far better choice.

things that have no known application, expands our thinking, and leads us to discoveries that we could not have imagined. Some of the most amazing things we have found are things we did not know we were looking for. This narrative can be compelling, but it is not often heard. Science as a whole would be better off if there were more public faces of science who were able to eloquently make this argument.

Narrative also plays a role in smaller scale decisions about support for science, sometimes down to the scale of the individual funding proposal. Because science funding in many countries follows the ebb and flow of politics, large funding initiatives tend to be linked to certain areas of inquiry, which change depending on what happens to be in fashion at the moment. This is particularly true in the United States, where every few years an emphasis is placed on some trendy topic. (Rarely if ever are these trends accompanied by a meaningful increase in funding). The narrative usually goes that these topics are key to maintaining global competitiveness, and green energy, or nanotechnology, or what-have-you is the key to success. The researchers who are most successful in securing funding are usually those who are most facile in their willingness to articulate how their work supports and advances the current hot topic, regardless of what they are working on.

Public policy, as we have said, is storytelling, and policy that provides funding is no exception.

16. Case Studies

DARTH VADER
The ability to destroy a planet is insignificant next to the power of The Force.

-STAR WARS (1977)

In this, our final chapter, we look at some of the largest and most visible public controversies surrounding science around the world today. In each case, our goal is to identify the predominant narrative proposition as most frequently viewed by people working in the sciences, articulate the competing narrative reference frame through which events are viewed by the opposition, and propose a possible counter-narrative that might be used to sway opinion in one direction or the other. As we discussed in Chapter Thirteen, the narrative reference frame to which someone subscribes is a function of a host of factors, and these frames are not easily abandoned. The advice contained herein is not intended to resolve any of these issues, but to demonstrate that there may be more effective strategies than those most commonly employed.

Case Study 1: GM Foods

Since the introduction of the first genetically modified food crops more than twenty years ago, controversy about them has raged in many parts of the world, particularly in Europe. People in many countries remain resistant to the introduction of these crops, despite the lack of evidence of a risk to human health. The two sides of this issue subscribe to two competing narrative reference frames.

In one telling, science is the hero, and what it wants is to bring health and prosperity to the world, but a poorly informed public is giving them trouble. Thanks to advances in genetic engineering, it is now possible to speed up the natural process of breeding that farmers have been using for millennia, producing crops that offer bigger yields, better resistance to pests, and many other advantages. As this new technology grows more and more sophisticated, it offers the potential to transform farming, and holds particular promise for the nations of the developing world. The obstacles standing in the way of progress are the obvious technological hurdles, and the meddlesome intervention of interest groups that stir up public fears of these crops despite a total lack of any evidence. Science is trying to provide a safe, abundant food supply through GM foods, and those who oppose them are standing in the way of progress.

In the alternate telling, farmers and consumers are the heroes, and what they want is to enjoy safe, healthy, natural foods, but science is trying to cram tainted food down their throats. Human beings have cultivated food crops for thousands of years, and those foods have been proven to be safe and beneficial. Now science, in service to the nefarious interests of government and agribusiness, has introduced foods that have been artificially concocted in a laboratory. These engineered foods rely on splicing together genes from other organisms, and as such constitute a foreign additive to the foods we have eaten for generations. Though the scientists claim that these foods are safe, they cannot be trusted, as they have a vested interest in getting us to eat this food. What's more, these crops are all new and we do not yet know what their long term effects might be. These tainted products should be aggressively restricted, and must be prominently labelled, if not banned altogether.

Notice how each of these two sides is neatly framed as a narrative proposition. In each case someone wants something badly, but has trouble getting it. As we noted in Chapter Thirteen, the fact that these two frames are diametrically opposed to one another means that two observers can have completely different interpretations of the same event. Imagine a major study is published showing that a certain GM crop has no negative health effects. One side will see the study as evidence that science is trying to improve the world and is being stymied by the opposition for no good reason. The other side will see yet another piece of propaganda from a morally compromised group with a hidden agenda. The two sides could shout these arguments at one another *ad infinitum*, but they would never get anywhere, because neither appreciates that what they think they are saying is not what their opponent is hearing.

A more effective approach would simply be to offer a counter-narrative. We can all agree that everyone wants safe, abundant, natural foods. Genetic modification is a new technology, but it may be moving us closer to that goal. We need to be careful, move slowly and deliberately, and do everything we can to make sure these foods are safe. Yet we should remember that GM foods may be healthier than conventional foods, if they enable cultivation without pesticides, or if they can be grown in regions of the world where hunger is a real problem. In order to address to challenges of a growing world population, we must consider all possible options, and we must weigh the risks of each new technology against its potential benefits. We all want a safe, healthy future, and it is by working together that we can best avoid the dangers along the way.

This counter narrative joins the two sides together, by emphasizing the thing they both want together. This larger, more inclusive narrative can embrace people from either position, and is far more likely to attract people from the undecided middle. By adopting a narrative proposition on which both sides can agree, we are able to shift the conversation away from what people *believe*, and begin a more substantive conversation about what society ought to *do*.

Case Study 2: Evolution

The theory of evolution has been the source of controversy ever since its introduction more than 150 years ago. Despite being universally embraced by all of science, despite the overwhelming mountain of evidence for the truth of the theory, despite the fact that it forms the backbone of all of modern biology, there are those who dispute its validity. Particularly in the United States, disbelief in evolution remains strong, and public controversies continue, particularly centered around whether the theory should be taught in public schools as the only explanation for life on earth, or taught alongside some alternative, non-scientific explanation. The sides of this debate fall into two easily distinguishable camps. In this case, the scientific theory offers a compelling narrative, while the opposition is able to construct an anti-science narrative about society.

The narrative around evolution goes something like this:

> All living things, from the least complex microorganism to the largest animal, wants to survive, and breed. Because there are so many living things on the earth competing for resources, and habitats, and because of the constant struggle of predator/prey interactions, the struggle to survive and breed is a difficult one. Over time, as organism reproduce, random mutations occur, creating an offspring that is slightly different than its parent. Sometimes these mutations create a slight competitive advantage for the offspring, for example better camouflage to hide from predators, or longer claws to ensnare prey. These advantages make the mutated offspring more likely to survive and thrive, which makes them more likely to have offspring of their own, and thus the trait is retained and becomes part of the species. In this way species

> change over time, and over billions of years this process has allowed life on earth to evolve from tiny archaea into the entire menagerie of species we see today.

This scientific narrative is usually accompanied by a policy narrative when it comes to education. In this narrative, society wants to teach its children the theory of evolution, because evolution is one of science's foundational ideas. Providing an education that adequately prepares young people to participate in the world means giving them some background in science, and it undermines that education to present an unscientific set of ideas alongside a well-established scientific fact.

The opposing narrative centers around the members of society who disagree with the theory. Their narrative goes something like this:

> We believe in a powerful and loving Creator, and we want to teach our children that they too are the product of a loving God. Political forces are using science as a weapon to persecute us, insisting that science proves that our Creator does not exist. We want the notion of creation to be presented in school, because we fear that allowing our children to be indoctrinated with ideas that deny the existence of our God will drive them away from our faith.

These two sides are entrenched, both are firmly convinced their view is the correct one, and the controversy spawns bitter disputes. Those proponents of each side whose views are most extreme are unlikely ever to be reconciled with one another. Some on the side of science have gone so far as to say that evolution firmly disproves the existence of God, while some in the creation camp insist that the planet and everything on it was formed from whole cloth some 8,000 years ago. People who subscribe to either of these notions are unlikely to find any common ground.

Yet there is a compelling counter narrative to be found, and it relies on the fact that the vast majority of the population holds neither of these extreme views. Certainly in America, the majority of the population believes that human beings were created by a divine being, and yet clear majorities also accept that species change over time. The solution then, is to embrace a narrative that allows for both ideas to exist. The narrative of evolution may in fact be true, that life wants to survive and breed, while at the same time, the reason it wants that is because that's what God inspires it to want. It may be true that genetic changes lead to new traits which are preserved as a result of competition, but it may also be true that those changes are not random events, but acts of God.

Many scientists balk at this notion, insisting that evolution is in some ways compromised or impure if it is presented as a mechanism that may be guided by the divine. Yet these scientists are perhaps overstating their own case. Much of the confusion, and thus the contention, around this issue stems from the use of the word *random*. Biologists refer to the "random" mutations that cause changes in species, by which they mean that the changes cannot be predicted, that they happen spontaneously, and we can see no force driving them.[1] Yet for most people, a "random" process is one which happens without method or direction, an unguided process, a process which is erratic, haphazard, and unconscious. Evolutionary biology doesn't suggest that gene mutations are any of these things, and it certainly doesn't prove that there is no God. Genetic mutations are the result of the quantum mechanical nature of biochemistry. If that is God's chosen instrument, even most scientists would agree that God works in mysterious ways.

Case Study 3: Climate Change

Our final example is by far the most contentious and most

1 This is, of course, not strictly true, we now know that there is a complex interaction between environment and gene expression that ultimately determines phenotype, but it is a valid first approximation.

consequential scientific issue in the world today, global climate change. The decisions that we make (or fail to make) over the next one or two generations will impact the lives of humans on this planet for the next thousand years. Climate change is a huge, sprawling issue with ramifications all over the globe. It is inextricably linked to geopolitics, to the global economy, to income inequality, and to the balance of power in the world. Dealing with climate change is a matter of international policy, and it is as much an economic question as it is an environmental one. The issue is as complex as it is vast. Here, we present an overview of the broad narratives embraced by the two sides of the climate change debate. Both narratives are powerful, because, as we discussed in Chapter Four, the stakes are incredibly high.

The pro-science narrative goes something like this:

> The climate is changing as a result of human activity. Those changes promise to have dire consequences for future generations, in the form of rising seas, worsening weather, droughts, crop failures, and more. We, the international community, must take action now, to change the way we live, reduce our consumption of fossil fuels and our emission of carbon dioxide, in order to avoid the worst of these consequences. The matter is urgent and the time for action is now, because if we do not change our habits soon, we will cross a critical threshold after which it will be too late. Those who doubt the science of climate of change or object to potential remedies are dangerous obstructionists who threaten the health of the planet itself. We must act now.

Powerful though the above narrative may be, progress addressing climate change around the world has been slow. Resistance is high, skepticism runs rampant. Some opponents of action have sowed the seeds of doubt, while others (particularly in the U.S.) have embraced outright denial and suppression of facts. This opposition can take several different narrative forms, but the most common (and probably the most convincing) goes something like this:

> The earth's climate is in a constant state of flux. Changes to the climate happen in a way that we have little ability to predict or control. A growing consensus of evidence now suggests that the earth's climate is currently undergoing one of these changes, but we cannot prove whether it is caused by human activity or not. Human societies are trying to grow, to bring prosperity and rising levels of wealth to more people around the world. Scientists, by exaggerating scant evidence that relies on computer simulations rather than actual experiments, threaten to undermine growth around the world. They propose limits on the use of resources that would harm the global economy, and condemn whole nations to perpetual poverty. We cannot choke off human productivity based solely on the overblown fears of a few intellectuals who offer no concrete proof of their claims.

Again, we see the entrenchment of the two sides. As we discussed in Chapter Thirteen, the divergence of these two frames means that two observers watching the same event can have completely opposite interpretations. If a major study is published showing results of modelling that suggest a certain limit on emissions required to stay below a certain threshold of temperature increase, one side will claim evidence for the urgency of action, while the other will claim yet another exaggerated and overblown piece of propaganda with no basis in fact. Neither side is likely to convince the other to abandon their position.

Instead, as in our previous examples, an alternative way forward is the construct a more inclusive narrative, one which embraces both sides. All of us want to preserve the health of the planet, all of us want a secure future for our children and grandchildren. But all of us must admit that at present, almost every facet of modern life depends on fossil fuels, and changing our patterns of consumption will be complicated and costly. What's more, there is a tremendous amount of uncertainty around climate change. We are certain the climate is changing. We are nearly certain the change is the result of human activity. But we are far less certain that the interventions we propose will have the effect we predict. There is no historical precedent for changes of this magnitude, so we must rely on numerical models without any way to test their validity. The truth is that it may already be too late. We cannot blithely assume that reducing our carbon dioxide output will solve the problems associated with climate change.

Again, here we are constructing a larger, more inclusive narrative. We are constructing a narrative proposition that joins us all together as one protagonist, by establishing a goal that we all have in common. This narrative invites people in rather than driving them out, because it doesn't require them to abandon their pre-existing notions. Instead it establishes something we all want badly, and shifts the focus onto the trouble we are having getting it. In this way, we can stop arguing about what one another believes, and begin a reasoned conversation about what we are going to *do*.

Scientific controversies are not easy, and they are not new. They have been with us since the dawn of science. The case studies presented in this chapter are meant to serve as examples. If these controversies are someday settled, new ones will emerge in their place. When they do, one can be certain that both sides will view events through a certain narrative reference frame, and that the path toward resolution will involve finding another, larger frame that allows the opposing views to coexist.

Afterword

On the north side of the city of Los Angeles, there is an iconic white building called the Griffith Observatory. Built in the mid-1930s, the observatory's distinctive Art Deco architecture and prominent location on a hill overlooking the city have made it one of Los Angeles' most recognizable landmarks, and it is free and open to the public, in accordance with its mission to make astronomy accesible to people of all ages.

Needless to say, I love the place, and during my second stint in graduate school I chose it as a location for one of my student films. One afternoon while scouting the location with two of my classmates, my cinematographer commented on the moon, which was two days past its first quarter, and was visible above the eastern horizon. "It'll look better at the end of the week," I told him, "when it will be full, and will rise as the sun sets, over there." I pointed eastward to the distant ridges of the San Gabriel mountains.

"How do you know that?" he asked, sounding astonished. I forget exactly how I answered him, probably mumbling something about my undergrad days. But what I should have said, what I wish had said, is: "Because I've watched it, and it does the same thing every time." I thought about that moment often over the course of writing this book, because I think it says something about the relationship between humans today and the natural world.

For most of human history, people were intimately familiar with the movement of the heavens. The cycle of the sun and moon, the apparent movement of the sun across the constellations of the zodiac, and the behavior of the wandering stars called planets were well known to peoples across the world, regardless of their culture. In the days before electric lighting, when a truly dark sky could be seen almost anywhere on earth, the patterns of the cosmos played a central role in people's lives. The heavens were a calendar that told of the coming of the seasons, and marked the passage of years. And so it is no wonder that in many cultures, peoples came to believe that the things they saw in the night sky controlled events here on earth. In their attempt to understand those events, and pass knowledge on from one generation to the next, people told stories about the heavens, and in many cultures those stories involved how we as a people came to live where we did, and what became of us when our time in this world ended. Many of the most important religions and philosophies of today have some origin in humankind's attempt to make sense of the cosmos, and science is no exception.

Some people think of science as the search for truth, but truth, like beauty, is in the eye of the beholder. I prefer to think of science as a story that is constantly being re-written, a story we tell that tries to make sense of not just the heavens, but of everything we observe and experience in the universe. Yet unlike the stories told in religious traditions, the tenets of science are universal, they apply to everyone, everywhere, and they are verifiable; science is a story that stands up to scrutiny. That story has been one of the most influential in the history of our species, and it has made the world we see today.

Yet for all its power, science is not without its weaknesses. Science tells us so much about the world around us, but so little about our place in that world. There is a short list of existential questions that almost every human culture has grappled with, and in most cases, the answers that science provides are either terribly unsatisfying or nonexistent. Science tells a story of a universe that abruptly began in a massive explosion, a species that emerged by chance, and a conciousness that disappears the instant we die. It has nothing to say about love, or charity, or kindness, and it cannot tell us the difference between right and wrong.

When I hear (as I often do) friends or colleagues bemoaning a general lack of interest or understanding in a certain topic in science, I think about my cinematographer classmate and that afternoon at the Griffith Observatory. He was a smart and well educated guy, a genuinely curious person. But he knew nothing about the most basic cycle of the phases of the moon. He was completely ignorant of a thing that his ancestors had known for thousands of years. And it wasn't because he wasn't interested or because he didn't care. He didn't know it because he'd never *seen* it.

Science and technology have shaped the world we live in, but in so doing they have insulated us from many of the experiences of the world that caused us to be curious in the first place. The day may come when science ends, when eveything that we want to know about the universe is known.[1] I worry sometimes that if and when that day comes, we will have built a world so isolated that we will no longer have any contact with the universe we know so much about. I think part of the apathy or disregard for science we see in society today is a result of this isolation. My cinematographer classmate lived in a society where people lived indoors and everyone had electric lighting, and thus knew nothing about the moon. The more things science gives us, the less we seem to care.

If I could change one thing about the way we teach and present science around the world, I would have us stop trying to give people answers, and start trying to give people questions. The real beauty of science, the excitment, lies not in what is known, but in what is unknown. The way to inspire someone to pursue science is not give them knowledge, but to make them *desire* it. Each of us who was drawn to the sciences wanted something badly, and each of us had our own trouble getting it. If you want to get

1 Scholars disagree about this question. One of the best explorations of it to date can be found in *The Island of Knowledge*, by Marcelo Gleiser.

someone excited about science, give them something to want.

Once they want it badly, they will find the trouble on their own.

-Tim Miller
October, 2015

Acknowledgements

This book would not have been possible without the support and encouragement of my friends, family, and close colleagues. I need to thank some of the great teachers I have known or worked with, including: Joe Sutton, John Thorstensen, Marcelo Gleiser, Francis Kennedy, Eric Mazur, George Whitesides, Bob Westervelt, Tom Abrams, Norm Hollyn, Tom Holman, and Nevin Schreiner. I owe a huge debt to the great people I know in the informal science education community, including: Larry Bell, Paul Martin, Troy Livingston, Eric Seigel, Rob Semper, Tom Rockwell, Rae Ostman, Catherine McCarthy, Brad Herring, and many more. I am grateful to my many clients, and most especially to the National Science Foundation, who has provided the lion's share of the funding for most of the projects I have worked on over the years. I owe a tremendous debt to Alan Alda, whose work continues to inspire me, and to the entire staff at the Alan Alda Center for Communicating Science at Stony Brook University, especially Christine O'Connell, Valerie Lanz-Gefroh, Lydia Hodges, Steve Marsh, Elizabeth Bass, and Howard Schneider. I could not have done this without Tom Breen, who showed tremendous patience with me during the editing process, or without Sadie Bride, who makes it all possible.

About the Author

Tim Miller is a science communications consultant and an instructor at the Alan Alda Center for Communicating Science at Stony Brook University. A former education associate at the Boston Museum of Science, and research associate at Harvard University, he has given lectures and workshops on science communication for universities, government agencies, and professional societies across the U.S. He holds degrees in physics and engineering from Dartmouth College, and is an alumnus of the School of Cinematic Arts at the University of Southern California. He lives in Connecticut.

Made in the USA
Monee, IL
30 March 2026

47326441R00125